Implementing CHANGE Through LEARNING

To those early pioneers, intellectually endowed and realistically wise, who first conceptualized the concepts of school change and how it could be guided by strategic actions and facilitated; who created the ideas of Stages of Concern, Levels of Use, and Innovation Configuration . . .

and to the nation's educational funding agencies who chose to support rigorous research to develop the knowledge base for employing these ideas for the improvement of our nation's schools and school systems . . .

and to those believers who used these research findings, added their experiences and modest modifications—all of which have provided a set of understandings and insights about how to go about improving organizational effectiveness especially for our schools, we dedicate this volume to you.

Implementing CHANGE Through LEARNING

Concerns-Based Concepts, Tools, and Strategies for Guiding Change

Shirley M. Hord • James L. Roussin

Foreword and Finale by Gene E. Hall

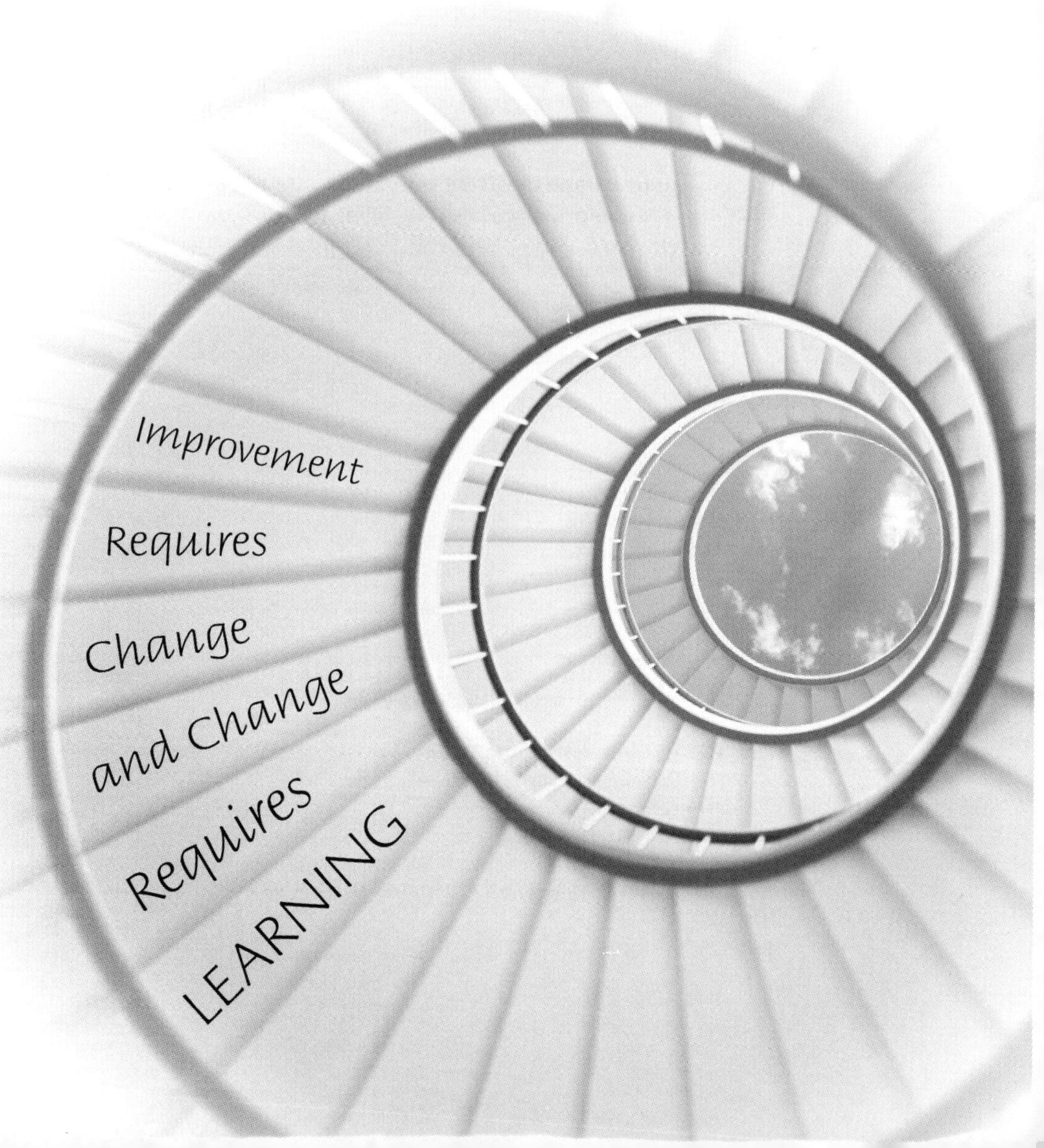

SAGE | AdvantEDGE Teacher Resources

Los Angeles | London | New Delhi
Singapore | Washington DC | Melbourne

Originally published in 2013 by

CORWIN
A SAGE Company

Corwin
A SAGE Company
2455 Teller Road
Thousand Oaks, CA 91320
www.corwin.com

This SAGE AdvantEDGE edition published in 2017 by

SAGE | AdvantEDGE Teacher Resources

SAGE Publications India Pvt Ltd
B1/I-1 Mohan Cooperative Industrial Area
Mathura Road
New Delhi 110 044, India

Published by Vivek Mehra for SAGE Publications India Pvt Ltd and printed at Chaman Printers Pvt Ltd, New Delhi.

ISBN: 978-93-860-6227-7 (PB)

Contents

Foreword

You have picked up an unusual and special book. Your reason(s) for selecting this book at this time probably is related to your being engaged with implementation of a particular program, innovation bundle, or a system change such as developing a professional learning community. Or perhaps you are mainly interested in seeing how research-verified constructs and tools can be used to facilitate change processes. An additional reason may be your knowing one or more of the authors. Whatever your reason, you are about to explore a book that is unusual (in regard to its content and utility) and special (in terms of the authors' expertise and passion).

When Shirley asked me to write a foreword, she observed that today the authors of many forewords add other ideas. "They don't just stick to providing an introduction to the book." This is an approach different from what I am used to. So I will do a little of both by introducing a little of what you will be learning about and sticking in my two cents' worth.

You will learn a little about each author in their bios. The unusual organization of the book begins with two atypical pieces, the authors' beliefs and a section that invites you to imagine how education could be. These will tell you more about how the authors view leadership and the significance of understanding the importance of *learning* as change processes unfold. These pieces also begin to make this book personal, which is an important component of having change processes become successful. Knowing the people and understanding their feelings is a critical component of change leadership.

The main topic of this book is offering actions that can be used to facilitate and implement change. In each chapter, constructs and tools of the Concerns-Based Adoption Model (CBAM; Hall & Hord, 2011) provide the conceptual and diagnostic framework for understanding the dynamics within change processes. These same CBAM constructs and tools then become the diagnostic bases for determining what to do. Each chapter goes beyond theory by offering concrete examples of actions that can be used to facilitate implementation.

For those of you who may not know about CBAM, it is a perspective for understanding change that was first proposed as a framework by Hall, Wallace, and Dossett in 1973. Rather than beginning with viewing change at the system level, the foundation of CBAM is developing understanding of how change is experienced by individual implementers (e.g., teachers, school leaders). Their feelings and perceptions (Stages of Concern), their gradual development of expertise (Levels of Use), and the extent to which they use the innovation with fidelity (Innovation Configurations) are the three primary constructs for assessing implementation. From this foundation the CBAM perspectives build to address the importance of leadership and system context. These three *diagnostic dimensions* are the evidence-based tools that are applied so well in this text.

THE PLC AS CONTEXT

An important foundational piece and place to begin is to provide a little of my understanding about what a professional learning community (PLC) is and is not. (I should note that I see a PLC as another innovation that has to be implemented; it doesn't just spontaneously arrive in any organization.) A related reason I have chosen to refer to PLCs at this point is, in part, to acknowledge Shirley's strong beliefs about what a PLC can be and also to use it as a metaphor of the approach the author team has taken in writing this book. I see a PLC as being a high-level configuration of organization culture. It is about the cultural norms, beliefs, and values that are shared; the types of information and ideas that are talked about; and the extent to which there is a shared vision and team approach to accomplishing the mission.

In many ways the authors of this book have functioned as a PLC. They share beliefs in the critical importance of schools and that learning for kids *and* adults is the primary aim. Each author has assumed major responsibility for creation and development of selected tasks. Each author has provided critiques of each other's drafts, and the feedback has been accepted as intended to improve the product, not as a criticism of the person. In addition, throughout their careers each of the authors has lived in the trenches with various types of change efforts.

In schools and school districts, a PLC exists when not only students but also the adults are engaged in learning *all the time.* I have spent time in many schools where the adults say something like, "Our PLCs meet at 8:30 on Fridays." As soon as I hear this, I know that the configuration of PLC in that school does not match my ideal.

Meeting at 8:30 on Fridays is just another required meeting. It is a structural piece only. What I want to hear is how throughout the week the adults exchange ideas, practices, and reflections related to what they are learning and what their students are learning. In my view the ideal PLC is what Peter Senge (1990) has called a *learning organization.* Ideas are exchanged, openly discussed, and disagreements are not taken personally. There is continual experimentation and mutual interest in what can be discovered, shared, and used again. If you could review the emails related to writing this book, you would see each of these indicators in action.

From a change process perspective, there is another understanding about development of an ideal PLC that I see as being important. If a PLC is a particular configuration of organization culture, then developing this type of culture requires understanding how organization cultures are shaped. I am bothered when I hear superintendents, policymakers, and, yes, professors direct principals to "change the culture in your school." I do not see that leaders have the power to unilaterally change culture.

Instead, development of an organization culture is a *social construction*. It isn't just what a leader does; it is much more related to how the participants in the school give meaning to the actions of the leader and others. It is through the adults sharing their interpretations of meaning that organization culture is shaped. No matter what the intentions of the leader may be, others will construct their own version of what these mean. For example, the principal may have been well intentioned in calling a special staff meeting to talk about test scores. The principal's intention could have been to celebrate growth. But some of the staff may interpret the called meeting as an attempt to change what teachers do. "You know what happened last time. All we heard about was how terrible the lunch room looked."

THE AUTHOR TEAM

Construction and implementation of an ideal PLC is not accomplished easily or quickly. Ideas, tools, and activities to help in achieving such a worthwhile aim is what this book is about. For more than a decade Shirley has been crisscrossing the globe as the champion of the PLC movement. I mean this literally; she has been not only all around the United States, but as far away as China. She truly sees schools being much better for kids and the adults when there is a PLC-type organization culture. Her leadership of the author team for this book reflects this passion. She has been the leader, she has listened, and at the same time she has pushed. The authors have shared ideas, debated, and come to consensus, and each has learned.

Jim's contributions to this book are equally grounded in the realities of what it takes to bring about change. He has been a longtime collaborator around CBAM work. He brings clear examples and innovative suggestions for how to use CBAM constructs and tools to facilitate implementation. As Shirley observed, "Jim is a voracious reader (and remembers what he has read) and is a deep and reflective thinker."

I may have gone on too long here about the importance of PLCs in general and how the author team has functioned, but you get to learn from the final product. Many good and useful ideas are introduced in this book, and they represent grounded ways to advance change processes.

If I may have one more page of indulgence, I would like to share one of my favorite metaphors. I first proposed this metaphor in 1999 in a theme issue of the *Journal of Classroom Interaction*. We had done a major study about the implementation of the then innovative National Council of Teachers of Mathematics (NCTM) standards. The study was done in the Hessen (Germany) School District of U.S. DoDEA. As you may know, the NCTM standards required a transformative paradigm shift on the part of teachers and administrators. To help illustrate the enormity of the expected change and to illustrate how CBAM constructs could be applied, I proposed the Implementation Bridge.

The Implementation Bridge symbolizes much of what is addressed in this book. Change is a process, not an event, which means having to journey across the bridge. Given the size of most change efforts, implementation entails more than simply jumping across a crack in the sidewalk. If the change is

The Implementation Bridge

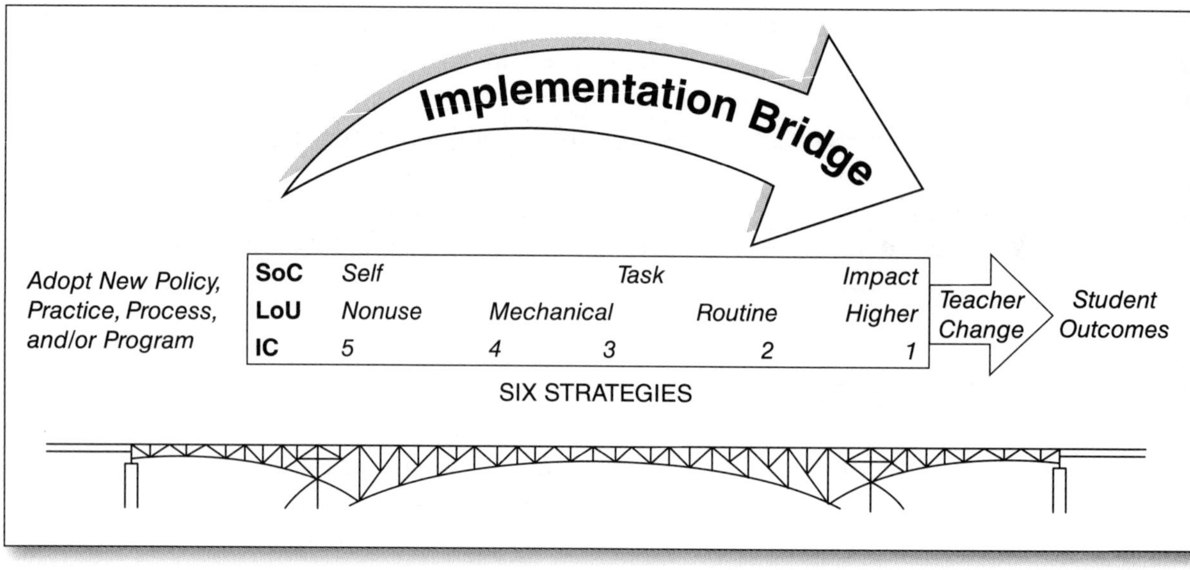

Source: Adapted by James Roussin from Hall (1999).

transformative, such as the paradigm shift from being teaching centered to student learning centered, the journey will take even longer. In other words, Implementation Bridges come in different lengths. Getting across major chasms takes time and ongoing support. There is, of course, no need to conduct a summative evaluation until implementers are across the bridge.

But formative evaluation is very important as people are approaching and moving across the bridge. In CBAM parlance we conduct *implementation assessments.* We assess Stages of Concern, Levels of Use, and Innovation Configurations at regular intervals. An Implementation Assessment Report is then presented to the change leaders. This report not only summarizes the current extent of implementation (i.e., how far across the bridge each implementer has progressed), it will also provide recommendations for what should be done next to facilitate getting implementers farther across the bridge.

What you will discover in reading this book is that in each chapter each of the three CBAM tools is used to assess the current extent of implementation. The same information is then used to introduce concrete examples of actions that can be taken to further advance implementation. In other words, each chapter describes examples of how to assess where implementers currently are on the bridge and provides examples of interventions that could be used to facilitate their moving farther across the bridge.

Shirley and Jim worked hard to bring to you what they have learned about implementing change. I think their ideas will be of interest and help no matter the size of your Implementation Bridge. What I particularly like about this book is the way each chapter illustrates how research-verified change constructs and tools can be used in practical ways to facilitate implementation. This really is a how-to manual.

Gene E. Hall

Preface: What You Will Find in This Book

The purpose of this book is to guide educational change leaders through a concise, step-by-step process of change implementation over time in order to ensure success and, as a result, build professional capacity through that effort. The bottom line is that implementation is a necessity—an imperative—for any successful change, and it has more often than not been overlooked in most school change efforts.

There is a great deal we already know about change. So this book is not about new theories or recent research. It identifies what we believe are the available, tested, research-based *best practices* in bringing forth any endeavor of change toward an end that is results oriented and guided by human interaction and vision. While there is a lot we already know, much of that knowledge is not applied in many, if not most, change initiatives.

Today, more than ever, we are urgently in need of success in our change projects. As we enter the 21st century, we are on the cusp of significant educational reform opportunities that will radically change how we do business in education. What does this mean for educators? While change has been a continuous topic of conversation for most professional educators for many decades, this century is going to require teachers and administrators to totally rethink how education is delivered to children. This is going to require significant changes across all sectors of public education.

Our current education system has been built around learning conducted mostly in classrooms, from textbooks, and from individual teachers. The future will look much different, as noted at the U.S. Department of Education website.

The challenge for our education system is to leverage technology to create relevant learning experiences that mirror students' daily lives and the reality of their futures. We live in a highly mobile, globally connected society in which young Americans will have more jobs and more careers in their lifetimes than their parents did. Learning can no longer be confined to the years we spend in school or the hours we spend in the classroom: It must be life-long, life-wide, and available on demand (Bransford et al., 2006; U.S. Department of Education, 2010).

All of this change is going to require that educational leaders at every level become highly proficient and savvy as change leaders!

What should you, the reader, expect to find in this book?

Chapter 1: Introduction

Chapter 1 shares the philosophies and approaches that we have used throughout decades of work in improving schools through change and learning efforts. We share our beliefs about change so that openness, candor, and transparency are employed. And, importantly, Chapter 1 contains an expression of the value of the continuous exploration of and learning about change process that we have engaged in for our professional lives. How we view the importance of adult learning in these efforts is no secret and will become abundantly clear.

CBAM Components

Four chapters follow, and these are devoted to teaching about the components of the CBAM. Chapter 2 highlights strategies that change leaders use to reach success. Chapter 3 provides the vision of the change. Chapter 4 enables us to understand individuals. And Chapter 5 describes the behaviors of implementers.

This volume of well-developed, user-friendly, and clearly articulated Learning Maps has been created to provide guidance to change leaders in their learning about how to achieve successful change. Each Learning Map requires 1- to 2-hour time slots, with all materials supplied. This book has been created so pages can be reproduced by the individual who is leading or assisting in the change journey. Each Learning Map has a clearly articulated result or outcome of the event, in addition to clearly numbered steps to guide the learning group in the activity. These learning paths could be used by individuals for independent self-directed learning, but we know that learning done in a social context results in richer, deeper understanding and meaning. Thus, we recommend that at least several change leaders and their participants come to learn, together.

The focus of each learning map is to provide clear instructions as well as application to the strategies (Chapter 2) and the concepts and their tools (Chapters 3–5) in guiding implementation. These tools are especially helpful in addressing the needs of implementers who are experiencing widely divergent progress in their change efforts.

In an effort to make a highly complex topic a trifle easier to understand, and as an organizing framework, we use the story of a fictitious school district and its staff as we employ the six strategies in the district's change projects. The change story provides the district context (size, socioeconomic status, student demographics, significant factors of staff, central office organization, school board posture, etc.) of EveryWhere District. The change story captures the challenges any change leader has to face in bringing about successful change. The story also foreshadows the rationale for successful change by using the appropriate strategies, concepts, and tools of CBAM to assist in the desired change in EveryWhere District.

Chapter 2: Six Strategies

Moving From Adoption to Full Implementation

Chapter 2 focuses on the following six strategies that are part of the CBAM suite of concepts, tools, and measures for ensuring that change results in successful implementation. These actions (that are the responsibility of change leaders or facilitators) can serve as the initial guide and benchmarks for the journey of the change process in any organization.

1. Creating a shared vision of the change
2. Planning and identifying resources necessary for the change
3. Investing in professional development/professional learning
4. Checking or assessing progress
5. Providing assistance
6. Creating a context conducive to change

The research on staff development (Joyce & Showers, 2002) parallels school change research (Hall & Hord, 2011) in terms of how implementers learn what the new practice is and how to use it. The results of staff development research instruct us that investing in large-group learning sessions ("investing in professional development") must be followed by individual and small-group interactions ("checking or assessing progress" and "providing assistance") to clarify, correct, and enable implementers to continuously refine their learning. This assessing and assisting is the basis for what coaches and change leaders do in supporting educators to change, and both school change and staff development research are very clear about the imperative of such support for individuals in their learning and change efforts. The Learning Maps in Chapter 2 teach what the strategies are and how to use them productively.

Chapter 3: Innovation Configurations

Creating a Vision of the Change

However, the six strategies are not applied in isolation of understanding what the change precisely is and where the implementers are in terms of their reactions and their behaviors related to the change. CBAM provides significant tools that inform us about the implementing individual and how she or he constructs and conducts the strategic work of the change effort. The first of these tools for consideration is Innovation Configurations. The creation of a map represents both the ideal and the various ways that implementers will implement an innovation (as they are learning how to use it). It serves as an excellent tool for establishing expectations for what the user will do, but also the map identifies where the implementer is currently operating along the continuum of the variations. This identification is the basis for creating support and assistance to help the implementer learn more about using the innovation and moving ultimately to *ideal use,* the goal for the change effort. The Learning Maps in Chapter 3 help us guide change facilitators in doing this.

Chapter 4: Stages of Concern

Understanding Individuals

Through many years of research and its application in schools where change and improvement is the goal, we know all too well that individuals, for a variety of reasons, become involved in change in a wide variety of ways. Understanding where the individual is coming from, and his or her attitudes and reactions to change, is essential to formulating the necessary support to encourage progress in learning about the change and in using it productively. Working sensitively with individuals, identifying and understanding their concerns, and using that information positively are the outcomes of this chapter's Learning Maps.

Chapter 5: Levels of Use

Using Innovations

While Stages of Concern (SoC) deal with the affective side of change, Levels of Use (LoU) describe profiles of behaviors exhibited by implementers. This is a second concept and tool for understanding the individual in the process of change. Many change leaders prefer, if there are resources for using only one tool that focuses on the individual, to employ LoU, for it represents observable activities, therefore making it more readily understandable for some change leaders. Like the other Learning Maps, the ones in Chapter 5 focus on enabling the change leader to understand more precisely how the implementer is interacting with the innovation. Appropriate support based on this understanding adds significantly to the probability that the change effort will be successful.

EveryWhere's Journey Comes to an End, or Does It?

The story of EveryWhere School District, which has been used to introduce, explain, and suggest applications of the CBAM concepts, strategies, tools, and techniques, draws the curtains on this 3-year district change drama. The story has illustrated how a district change project has been initiated with "lessons" about the change process and how its ideas, theories, and tools for practitioners can be learned, developed, and applied.

We strongly urge participants using the Learning Maps to create a notebook in which to keep these lessons and their related ideas and reflections, to reserve a space for journaling. In this way, new ideas, fresh insights, and challenging commentaries may find a home for future perusal.

Finally, we have created the following chart to indicate the topic and outcome of each of the lessons that we are naming Learning Maps, to provide the learners who are using this book with an easy reference to the focus of each map.

Learning Map Reference Chart

Learning Map	*Topic*	*Outcome*
Learning Map 2.1	Explaining Six Research-Based Strategies for Change	Learners identify the six research-based strategies for change and explain why they are required.
Learning Map 2.2	Planning Strategies for a Change Effort	Learners create initial plans for a change effort that is focused on the strategies, and they explain how these will be used to cross the implementation bridge.
Learning Map 2.3	Reviewing the Literature on Structural and Relational Conditions for Change	Learners briefly describe a selected set of contextual factors, accessed from the literature, that are valued for successfully introducing changes in organizations (schools and districts).
Learning Map 2.4	Assessing Change Readiness	Learners describe five change readiness dimensions for determining staff willingness and capacity to participate in implementing a change.
Learning Map 3.1	Articulating the Need for Precision About the Change	Learners explain the imperative for creating a mental image—a written picture—of the change when it is in operation.
Learning Map 3.2	Identifying Structures of an Innovation Configuration Map	Learners identify and define the two major structures of an Innovation Configuration (IC) map.
Learning Map 3.3	Creating an IC Map With Guided Practice	Learners produce an IC map in a collaborative, guided-practice setting.
Learning Map 3.4	Developing an IC Map Independently	Learners produce an IC map of their change, using the skills developed from the previous sessions and working with a small collaborative group.
Learning Map 3.5	Reviewing and Revising the Map	Participants produce a reviewed and revised edition of their developing IC map.
Learning Map 3.6	Field-Testing and Revising the Map	Learners produce a field-tested and revised map.
Learning Map 3.7	Sharing the Map With Implementers	Learners create a plan for sharing the IC map with implementers.
Learning Map 3.8	Using an IC Map for Developing an Implementation Plan	Learners identify how the IC map can be used to initiate planning for implementing the change.

(Continued)

(Continued)

Learning Map	*Topic*	*Outcome*
Learning Map 4.1	Considering the Compelling Case for Concerns	Learners explain the concept of Stages of Concern (SoC) and use an individual's commentary to identify his or her concerns.
Learning Map 4.2	Generating Responses to Concerns	Learners suggest assistance and support appropriate to each individual's SoC.
Learning Map 4.3	Collecting Concerns Data	Learners describe two methods for collecting SoC data and match the appropriate method to a specific purpose; learners identify a third data collection method and its purpose.
Learning Map 5.1	Articulating Behaviors Associated With the Use of Innovations	Learners describe eight specific behaviors associated with an individual's learning to use innovations.
Learning Map 5.2	Identifying the Level of Use of Individuals	Learners identify an individual's Level of Use (LoU) and suggest appropriate responses to the identified LoU.
Learning Map 5.3	Collecting Levels-of-Use Data	Learners conduct informal LoU interviews to collect data about implementers and identify the individual's LoU.

Acknowledgments

Our profound thanks to Gene for the Foreword and Finale, and for his wise leadership in the field of significant school change. For the Concerns-Based Adoption Model, he has been the primary architect and stalwart caretaker of this body of work that has supported the change and improvement efforts of so many around the globe for many years. Further, he was an early model of the collaborative leader, inviting the voices of the research staff and school personnel to the conversation, planning, and decision making. In this way he has promoted leadership in others and the increasing professionalism of all.

Our sincere thanks go to professional development specialists and supporters of adult learning Dr. Edward Tobia and Terry Morganti-Fisher for bravely reviewing our first proposal to provide feedback on our initial thinking.

To Dr. William A. (Bill) Sommers, thoughtful reader and never-ending learner, you have our everlasting thanks for reading the initial complete draft to tell us where we were okay and where we faltered.

And to principals Dr. Nancy Flynn, of Ramsey Junior High School, and Barbara Evangelist, of St. Paul Music Academy (both of St. Paul Public Schools), who have demonstrated a keener understanding for what it means to be a collegial facilitator in guiding successful change across the school—and from whom we have learned—we give our warm thanks.

For the energetic, enthusiastic, and constructively critical reviewers of the production draft, please accept our boundless thanks for your careful read, thoughtful commentaries, and wise reflections that have immeasurably contributed to this manuscript.

It is impossible not to express abundant gratitude for the guidance of Dan Alpert, premier acquisitions editor for Corwin, for his belief in the process and product of this project. His critique and support of its development has been endlessly valued.

PUBLISHER'S ACKNOWLEDGMENTS

Corwin gratefully acknowledges the contributions of the following reviewers:

Kenneth Arndt
Professor of Educational Leadership, Argosy University
Schaumburg, IL

Sally Bennett
Superintendent, Armorel School District
Armorel, AR

Gary Bloom
Superintendent, Santa Cruz City Schools
Soquel, CA

Dalane E. Bouillion
Associate Superintendent for Curriculum and Instruction, Spring Independent School District
Houston, TX

Janice Bradley
Mathematics Leadership Coordinator, New Mexico State University
Las Cruces, NM

Kate A. Foley
Associate Superintendent, Naperville Community School District 203
Naperville, IL

Delores Lindsey
Professor of Education and Education Consultant, California State University, San Marcos
Escondido, CA

Terry Morganti-Fisher
Educational Consultant, Learning Forward and QLD Learning
Austin, TX

Jennifer Ramamoorthi
Teacher, Community Consolidated School District 21
Wheeling, IL

Judith Rogers
K–5 Mathematics Specialist, Tucson Unified School District
Tucson, AZ

Adrienne Tedesco
Instructional Coach, Gwinnett County Public Schools
Suwanee, GA

Claudia Thompson
Academic Officer/Learning and Teaching, Peninsula School District
Gig Harbor, WA

Edward F. Tobia
Project Director, SEDL
Austin, TX

About the Authors

Shirley M. Hord, PhD, is the scholar laureate of Learning Forward (previously the National Staff Development Council), following her retirement as Scholar Emerita at the Southwest Educational Development Laboratory in Austin, Texas. There she directed the Strategies for Increasing Student Success Program. She continues to design and coordinate professional development activities related to educational change and improvement, school leadership, and the creation of professional learning communities.

Her early roles as elementary school classroom teacher and university science education faculty at The University of Texas at Austin were followed by her appointment as codirector of Research on the Improvement Process at the Research and Development Center for Teacher Education at The University of Texas at Austin. There she administered and conducted research on school improvement and the role of school leaders in school change.

She served as a fellow of the National Center for Effective Schools Research and Development and was U.S. representative to the Foundation for the International School Improvement Project, an international effort that develops research, training, and policy initiatives to support local school improvement practices.

In addition to working with educators at all levels across the United States and Canada, Shirley makes presentations and consults in Asia, Europe, Australia, Africa, and Mexico.

Her current interests focus on the creation and functioning of educational organizations as learning communities and the role of leaders who serve such organizations. She is the author of numerous articles and books, the most recent of which are *Implementing Change: Patterns, Principles, and Potholes* (3rd ed., with Gene E. Hall, 2011), *Reclaiming Our Teaching Profession: The Power of Educators Learning in Community* (with Edward F. Tobia, 2012), and *A Playbook for Professional Learning: Putting the Standards Into Action* (with Stephanie Hirsh, 2012).

James L. Roussin, MALS, has been committed to improving teaching and learning in schools across the United States and abroad throughout his professional career. He has worked as a language arts teacher; gifted coordinator; ESL coordinator; curriculum director; executive director of teaching, learning, and school improvement; adjunct professor; and educational consultant.

Jim is currently working as a strategic change consultant and is the executive director for Generative Learning (www.generative-learning.com).

Jim helped to revitalize the Minnesota Staff Development Council from 1998 to 2004 and served as its president for four of those years. He has also served as a trustee on the National Staff Development Council (now Learning Forward).

He is a teaching associate for Human Systems Dynamics (an institute that is using complexity theory to impact organizational development work). He is also a learning facilitator for Leadership Development, Cognitive Coaching, Adaptive Schools, QLD (Quality Leadership by Design)—S.M.A.R.T. Goals, and Program Evaluation.

In February 2006, Jim traveled to India on a Berkana Learning Journey to explore new forms of leadership that are emerging in global communities. And in 2009 he spent 4 months working in the Middle East with ASCD-Middle East in supporting the United Arab Emirates Ministry of Education in implementing new teacher development standards.

Jim's current interests focus on healthy organizations and human development through the lens of natural systems theory and complexity science.

Jim is the author of a variety of articles as well as a coauthor of the book *Guiding Professional Learning Communities: Inspiration, Challenge, Surprise, and Meaning* (with Shirley M. Hord and William A. Sommers, 2010).

Gene E. Hall, PhD, currently is professor of urban leadership at the University of Nevada, Las Vegas (UNLV). He began his academic career at the national R&D Center for Teacher Education at The University of Texas at Austin. This is where he, Shirley Hord, and other colleagues developed the Concerns-Based Adoption Model, which has been the career-long focus of his scholarship, teaching, and coaching of leaders.

Following many years at The University of Texas at Austin, Gene moved to the University of Florida, then to the University of Northern Colorado, and now UNLV. He has twice served as dean of the College of Education and throughout continued to study, lead, and coach change processes.

In addition to his work with education and business organizations across the United States, he has had a rich array of collaborative activities with colleagues in other countries, including Australia, Belgium, England, Taiwan, Hong Kong, The Netherlands, and New Zealand.

His current research is centered on the Three Change Facilitator Styles (initiators, managers, and responders) of leaders and their effects. These studies have documented strong relationships between teacher implementation success and principal leadership as well as between principal change facilitator style and student test scores. Recent publications include *Implementing Change: Patterns, Principles and Potholes* (3rd edition, with Shirley M. Hord), *Foundations of American Education* (with James A. Johnson, Diann Musial, and Donna M. Gollnick), and *Introduction to Teaching: Making a Difference in Student Learning* (with Linda F. Quinn and Donna M. Gollnick).

1

Introduction

Just Imagine . . . schools in the future where the drivers of change are those who are truly committed to their students' success and who are regularly monitoring the learning challenges and achievement of their students. Educators meet regularly in collaborative teams to discuss student data and reflect on what the data indicate about student results, and from their reflections, they carefully determine what actions can have the greatest impact to improve student learning. All improvements are conducted through the day-to-day flow of teaching and learning experiences, insights, and action research. Throughout the school there is the belief that everyone's perspective and understanding is valued, embraced, and considered in shaping a collective future for student success. These educators create their future and that of their students through social interactions and a shared collective vision.

Just Imagine . . . educators in the 21st century who *respect and embrace* change. They see each reform initiative as an opportunity for their own reflection, renewal, and growth. The pathway of change is their primary means to engage continuous learning and to increase professional capacity. There is diminished resistance to change. Multiple and continuous feedback loops inform everyone on a daily basis how the innovation (new practice) is moving forward and having an impact on adult and student learning. And these educators consider opposing ideas or perspectives as an opportunity to stretch thinking and provide different points of view.

They believe that real change is synonymous with learning. Just as Peter Senge (1990) has stated,

> Real learning gets to the heart of what it means to be human. Through learning we re-create ourselves. Through learning we become able to do something we never were able to do. Through learning we re-perceive the world and our relationship to it. Through learning we extend our capacity to create, to be part of the generative process of life. (p. 14)

Just Imagine . . . the role of educational leaders in the 21st century (principal, district office personnel, etc.) as conveners of change rather than the drivers. These positional leaders tend to the collegial, social fabric of school life and invite transformation through conversation, the exploration of new questions, and thoughtful action. They do this because they know change is not a static event. It is a dynamic process that emerges over time through meaningful interactions and the behaviors that follow the interactions. As conveners, leaders hold the understanding that for any change initiative to be successful, it cannot be seen as only a logistical process or an organizational mandate; it requires an understanding of human dynamics and how to engage individuals in social/relational space. The task, then, of these positional leaders is to create the social structures and opportunities that bring people together in order to deepen personal and collective meaning and find shared accountability for student learning. In many ways, these leaders are the social architects who engage what Kouzes and Posner (1995) call "the art of mobilizing others to want to struggle for shared aspirations" (p. 30).

Just Imagine . . . the implementation of all change is guided by professional conversation and coaching. Using the concepts of Stages of Concern, Levels of Use, and Innovation Configuration, and their associated strategies and tools, each person contributes by shaping a meaningful picture of the change and identifying appropriate adaptive strategies. Throughout the organization there is a culture of freedom that allows and works with ambivalence, dissonance, and uncertainty so everyone has the necessary time to evolve new understandings and capabilities related to the change. Coaching is widely available to support each staff member in reflection, thinking, and finding insights in order to move into greater competencies around the change. It is now well accepted that because we are fundamentally social beings, communication is the lubricant that makes the implementation sustainable and successful.

How we perceive changes is clearly impacted by our values and the beliefs from which they derive. We think it is important to share the following significant beliefs that we hold for change.

SIX BELIEFS ABOUT CHANGE

The "Imagine" scenario above is grounded in six beliefs we hold about successful change. These beliefs are the foundation for our thinking about how the implementation of change happens in its best possible way. We believe that . . .

All change is based on learning, and improvement is based on change. Learning is a critical component embedded in the change process. Learning enables people to discard past practices and find new behaviors appropriate for the innovation. At the center of all successful implementation of a change is the opportunity for adults to come together and learn. Unfortunately, there has been much change in the past driven by mundane data and hierarchal processes that often leave people feeling used and dispirited by the effort.

Implementing a change has greater success when it is guided through social interaction. Humans create meaning through interactions with each other and their environment—change is more sustainable when it is driven by conditions that invite people to engage in social learning. It is through our interactions with others that we not only create the future, but also make sense of the present and find new ways to take action. That is why . . .

Individuals have to change before the school can change. The starting point for implementing any innovation is the individual. Each person will have his or her own concerns around a change and can be at a different stage of readiness for adopting an innovation. Bandura (1997) reminds us that, for learning and development to take place (critical factors for any change), individuals have to exercise their own personal agency, the ability to influence oneself and his or her environment. That is why Stages of Concern, the foundational concept of the Concerns-Based Adoption Model (CBAM), is so important in facilitating successful implementation.

Change has an effect on the emotional and behavioral dimensions of humans. Change can be stressful and often leads people to feeling disoriented and confused. William Bridges (2009), in his book *Managing Transitions,* reminds us about the emotional dimension of change, and that by not addressing that aspect, many change initiatives fail. It is this dimension that is addressed by Stages of Concern, that identifies implementers' feelings and attitudes. It is important to understand also that when people are learning to use an innovation, they tend to move along a spectrum that ranges from no use to full use of the new practice, or program. The research-based concept of Levels of Use provides the change facilitator eight behavioral profiles to consider as individuals adopt and become more familiar with and more skilled in using an innovation. The Levels of Use, in concert with Stages of Concern, can be employed to lessen stress, support individuals, and provide authentic assistance to them.

People will more readily choose to change when they foresee how an innovation will enhance their work. If there were no relevant reason to change, why would a person choose to do something differently? When it comes to implementing an innovation, individuals should be able to envision how it will affect their day-to-day work and how the results of their work will become different and better. The Innovation Configuration Map invites a mental image of what the change will look like as it previews new behaviors.

A change leader's role is to facilitate the conversations that invite others to own the desired change. Successful implementation of a change will depend on the quality of conversations that invite personal and social investment. It is imperative that change leaders facilitate collegial interactions throughout the change process because we are fundamentally social beings. Conversations are not the end, but rather the means to gain action and envision new behaviors of the innovation. The CBAM was conceptualized and created with that end in mind. It gives direction to thoughtful and sensitive change facilitators as they engage in the challenging tasks of supporting others in adopting and implementing new practices for a desired change.

CRITICAL UNDERSTANDINGS

Many individuals equate the CBAM with Stages of Concern (SoC), one of the components of the CBAM, and its foundational concept. SoC, with the concepts of Levels of Use (LoU) and Innovation Configurations (IC) all noted modestly above, are typically used as a means for understanding what is occurring with an individual in the process of change and, subsequently, as the basis for creating appropriate support and assistance to the individual.

Change Process Strategies

What is not so well known is an additional, powerful component of the CBAM—a set of research-based strategies or activities, found as a result of rigorous, longitudinal studies, which guide and direct the necessary *actions* required for successful change. These strategies serve as the imperative for conducting change in an organization, and they enable us to see the big picture of a change effort. These strategies may be thought of as a game plan. Their purposes reflect the necessities for the success of change endeavors, and we share what we have learned about them through long-term observation and participation, and from intensive study and research.

Expanding the Change Leadership Roster

We make a point about collaborative leadership, for we see it as the appropriate model for today's ever-learning educational leader and a model for the staff as a continuous learner. Most certainly, a thoughtful, sensitive, and strong guide is necessary at the outset of a plan for change, but the wise and inclusive leader plans for and engages participants in continuous learning about the innovation and how to use it productively. This model of principal as "collegial facilitator" shares discussions of the following ideas:

- How are we doing?
- What is working and what is not?
- How do we do the challenging exploratory work of identifying needs for change and finding potential solutions?
- How can we support each other in learning about and implementing new practices/programs so that our teaching quality is increased and students learn more successfully?

We think this spells out the collegial facilitator's role, and guides and supports the development of the staff's collective efficacy. To that end, we expect to be your "whisper coach" who sits at your side, or on your shoulder, calling your attention to useful actions and strategies for moving your organization's change effort to its successful end, through the challenging efforts of the entire staff that is using analyzed CBAM data to ascertain the progress of implementation and to support its improvement.

AN INFORMAL INTRODUCTION

Although you will find our "scholarly" bios or vitae in the front pages of this book, we would like to say a word and introduce ourselves to you in this authors' note, in a much more informal manner.

Individually, we are blessed to have this remarkable team to share information, insights, and wisdom with you as a result of our participation with CBAM; we write in a rather familiar style, endeavoring to be straightforward and clear, but helpful. This is who we are:

- Jim, who is relatively new to writing about this topic, is a longtime user of CBAM in his work as a professional developer in colleges and schools. He is a voracious reader and is deeply thoughtful about the reading and always sensitive about his work with administrators and teachers in districts and schools. He is a fresh voice committed to the work that we all do to support educators to be the best that they can be. We are so pleased that he has encouraged this volume.
- Shirley has "been around for a long time," involved in the original studies and dedicated to adult learning as a means to increase the effectiveness of instruction and, subsequently, the successful learning of all students. She has been labeled as the "vigilante of PLCs" (professional learning communities), to which she responds, you're absolutely right! She discovered PLCs initially while working with Gene on the research work on CBAM.
- Gene, the wise and prophetic creator and architect of the CBAM, was director of the original research in a national research center, and champion of its results for schools. He maintains this role as an educational leader at the university level, while flying around the globe to continue dissemination of CBAM. He has been a mentor and friend and foremost a model of a collaborative leader, inviting staff's voice into the discussion and decision making about the research.

We have learned, through our experiences and the continuous study of what we do, that all change is based on learning, and improvement is based on change. Interestingly, we have learned also that for an individual (teacher, administrator, or other organizational member) to move from one SoC to the next, or from one LoU to a higher one, or to move closer to ideal practice on the IC Map, requires learning. This we will point out in the text—as a vital challenge to the adult learner, especially in schools.

To set the stage, we will introduce you to EveryWhere School District and Bob Yurself, whose actions will serve as the thread that pulls the material and its applications into stories and text that provide sense-making and understanding so you can use the ideas and constructs in your own change and improvement efforts.

EVERYWHERE SCHOOL DISTRICT

EveryWhere School District is located in a suburb not far from a large metropolitan center. In the last few years the district has experienced significant demographic changes with a large increase in English language learners and greater disparity in socioeconomic backgrounds of students across the district. And recently there has been a large turnover in staff due to retirements. A lot has changed in EveryWhere in the last decade.

For a long time, staff at EveryWhere considered themselves above average in serving children and preparing them to become successful, lifelong learners. However, recent data have painted a very different story. Principals have noted a significant increase in school suspensions, district data reveal that there is a larger number of dropouts than in the past, and recent state tests show that many students are struggling with basic mathematics skills compared to neighboring school districts. The staff at EveryWhere was surprised by these results and unsure what to do.

In order to resolve these challenges, the superintendent requested meetings with small groups of district and school staff members over a 1-month period to dialogue about what might be done to increase student success. At the end of the month, the most common suggestion by EveryWhere staff was to focus on one area for improvement. They suggested that the best place to begin was to identify needed changes in the district mathematics program and to commit to this effort for the next 3 to 5 years. The superintendent and school board agreed, knowing that educational research advocates for districtwide improvement goals that occur over time.

For the remainder of that year, staff members, parents, and students were invited to participate in small teams at various grade levels to gather math data, examine results, and determine where math learning was successful and where it was falling short. Some teams were asked to visit neighboring school districts to better understand what was accounting for their math success. Each month the board listened to what the teams were learning. To understand EveryWhere's challenges, and to find the right solutions, became a districtwide venture.

At the end of the year, EveryWhere staff could see specific changes that needed to occur in order to bring more success to their students. For example, the elementary mathematics textbook was not rigorous enough to prepare students for what they needed to know when entering the middle school. And the recently adopted math program in Grades 6–9 required students to be engaged in collaborative problem solving around algebraic concepts. This type of instruction was unfamiliar to what most of the math teachers had experienced in their professional careers. And at the high school level, math courses were now requiring the use of sophisticated technology instruments that teachers were not comfortable in using. EveryWhere staff could identify specific changes that would make a difference in teaching practices and, subsequently, for their students in learning mathematics.

EveryWhere was now emerging from the exploration and adoption stages of change to the implementation phase. It was agreed by everyone that the identified district changes would be implemented over the next three school years. In their reading about school change, the teams learned that successful implementation of change is a critical factor for seeing improved performance in student learning. EveryWhere staff wanted to do this change right. So it was decided to assign someone from the district as a change coach to facilitate and guide the implementation process. The superintendent asked Bob Yurself if he would serve in this role for the district.

Bob was the principal at the most successful school in the district and was highly respected by his staff. He had recently guided a curriculum adoption that made significant gains in reading at his school. Bob often believed that his success came because his staff was change savvy. Prior to the curriculum adoption, Bob facilitated a book study on change with the hope that his faculty would become more adaptive and resilient to change. They chose the book *Implementing Change: Patterns, Principles, and Potholes* (Hall & Hord, 2011). The staff went away from that learning opportunity understanding how critical implementation is for the success of any change process. Bob was excited to now be the district's change coach. He hoped to develop across the district the same change "savviness" that had been accomplished in his school.

We invite you to join Bob Yurself as he prepares EveryWhere School District for being change savvy in preparing to implement the new identified math innovations. You, along with Bob, will learn about successful implementation practices by using the set of change concepts and measures that is the CBAM. These concepts provide a systemic approach that describes, explains, and predicts teacher concerns and behaviors throughout any school change effort, in addition to identifying important strategies to be undertaken by a change coach.

2

Six Strategies

Moving From Adoption to Full Implementation

The superintendent informed Bob Yurself that he would be responsible for the coaching of campus staff across the district in the implementation of the three strategic change initiatives, in the elementary, middle, and high schools, to improve mathematics in each of the buildings across EveryWhere. Bob would be provided an assistant administrator for his school, so that half of his time could be devoted to his district-level coaching position for the change effort.

EVERYWHERE BEGINS THE IMPLEMENTATION JOURNEY

First, the elementary schools would adopt a new math textbook that would better prepare students for integrated math learning in the middle grades. The current elementary math text was not sufficiently comprehensive to prepare students, for they were now expected to do more than basic computation; they would learn to reason and communicate proficiently in mathematics.

Second, math teachers in Grades 6–9 did not understand the best way to facilitate instruction where students work together to solve problems, reason about possibilities, justify their ideas and solutions, and look for connections. The Connected Math Project (CMP) was quickly adopted to address state math scores, but without the appropriate professional development for teachers.

Because this was not the way the teachers had learned math, they were struggling to believe in the integrity of the program and were falling back on past practices. A strong recommendation of the CMP was that teachers interact in professional learning communities (PLCs) so they could address challenges, successes, and learn together. Bob would collaborate with the middle school math teams to implement the CMP through developing PLCs in the schools.

And third, teachers at the high school needed a 2-year professional development technology plan in place to ensure their skills in using the various technology tools that were now a mandatory component in most math courses. This would include the use of probes, graphing calculators, and scientific notebooks. This plan would increase teachers' professional capacity to understand how to use these technology tools, as well as embed them in the appropriate places in their curriculum to better support student learning.

Bob was excited to support the schools in making these identified changes. And while Bob was a highly effective change leader, he was also a master gardener. He knew that good seeds have a better chance of growing when the conditions are right and the soil is healthy. Actually, some master gardeners say the soil is the garden's palette. Just as the painter's palette holds the pigments the artist uses to create a masterpiece, soil provides the medium to make the garden grow and prosper. Bob saw a lot of parallels to gardening and school change, and it was this metaphor he often referenced in understanding how change happens.

The first thing Bob did was to create a change leadership team that had diverse representation from all levels of the schools. Together they would build a shared understanding for moving these changes forward. Bob would work with this team as a professional learning community. Each representative would be a learner of the innovations under adoption, and they would serve as coaches and facilitators for engaging others in their specific schools. Bob firmly believed that change is a process of cultivating relationships and that when people have the right conversations on topics that matter, it arouses passion, appropriate action, and shared accountability. Bob viewed his main job as facilitating the interactions of the team members so they, in turn, would engage others in co-creating the future focused on the expected changes. "People are the heart of all change" was Bob's motto.

As Bob thought about implementing these three significant changes, he returned to his metaphor of the garden. He remembered that growth would be enhanced or restrained by the soil, and the first step to good gardening is learning about the soil's composition (texture and structure) and how to improve it. Bob wanted to carry this idea over to the new math changes in preparing the soil for the upcoming change. He recalled the book he and his staff had read on change using the Concerns-Based Adoption Model (CBAM) and how six strategies were necessary for successful implementation of a change. They serve as the initial guide and benchmarks for the journey of the change.

1. Creating a shared vision of the change
2. Planning and identifying resources necessary for the change
3. Investing in professional development
4. Checking or assessing progress
5. Providing assistance
6. Creating a context conducive to change

EVERYWHERE'S FIRST LEARNING SESSIONS

Bob invited his change leadership team to engage in and interact around four learning opportunities to better understand the six strategies. He wanted his team to understand that there are research-based practices for change that make implementation more facile. For Bob, the strategies represent the bridge that makes it possible for the school and/or district to move from adoption to full implementation. And healthy soil represents an important part of the sixth strategy—the conditions of the context needed for bringing about successful implementation of the math changes.

And he wanted, along the way, to instill in his team the philosophy of sharing leadership so that it is distributed across grade levels and departments in the schools. He knew that when individuals have opportunities to make suggestions, discuss their perspectives, and engage in collaborative decision making, they become more invested in the changes under attention.

You (along with your change team) are invited to join Bob Yurself in four professional learning opportunities to better understand the six strategies that are necessary for successful implementation.

LEARNING MAP 2.1

Explaining Six Research-Based Strategies for Change

Outcome

Learners will identify the six research-based strategies for change and explain why they are required.

Assumption

Change does not happen simply because a change has been introduced; research can inform us about the steps to be taken to increase the potential for implementation of the change.

Suggested Time

60 minutes

Materials

1 copy, for each participant, of

Tobia, E. F., & Hord, S. M. (2002). Making the leap: Leadership, learning, and successful program implementation. *Instructional Leader.* Austin: Texas: Elementary Principals and Supervisors Association.

Handout 2.1, Assessing the Degree of Implementation of the Six Strategies

Engaging in Learning

1. Ask participants to organize themselves in groups of six—if there are not this many participants, arrange people in sense-making groups to do shared teaching/learning. Distribute a copy of the paper to each participant, directing each group to organize so that each person is responsible for one of the six strategies: to read and study, and plan to teach the others in the group about the strategy.

   Each person teaches others in the six-person group (please do not read the paper to your colleagues, but make a short learning plan that will involve the group members in the learning).

2. Reorganize the participants so that those who taught the same strategies meet in a group. In this group, they will each share how they taught the others the strategy. This group then makes a new plan to teach the entire group about their strategy, making use of the best ideas that they gained from the group. The group should select a person from their group to represent them and teach the large group of all participants about the strategy. Solicit questions and provide responses. This second "teaching" of the strategies reinforces the initial learning of the entire group.

3. Distribute Handout 2.1. Ask participants to identify a change that is in progress in their team, grade level, school, or district. Place the name of the change at the top of the handout. Review each strategy in the left column, and check the appropriate box to represent the degree to which the participant believes the strategy has been implemented in their change effort. In the box, jot evidence that supports your rating.

4. Prepare to report your findings to the large group: your surprises, certainties, questions, and curiosities. Solicit volunteers to report to the large group. Solicit questions and/or needs for clarification. Encourage responses/solutions from the large group.

   Remind participants of the time and location of the next session, and request that they bring their materials from Learning Map 2.1 to the next learning session.

MAKING THE LEAP

Leadership, Learning, and Successful Program Implementation

Edward F. Tobia, EdD
Shirley M. Hord, PhD

The public, the press, and the profession share a keen desire for their schools to become places where all students achieve to high standards. Over the last three decades many researchers have studied the efforts of schools to become more effective, so that students become more successful learners. Based on that research, many new programs have been introduced that hold the promise of improving student achievement. However, one significant finding relevant to this proliferation of programs is the lack of understanding about what should happen during implementation, i.e. the period of putting new programs and practices into place (Fullan, 2001).

Many educators, from policymakers to principals and classroom teachers, believe that if a school or district adopts a new program, somehow—after a workshop or two—the new program will be used effectively in classrooms and student learning results will improve in short measure. This "giant leap" in thinking overlooks the significant work that must occur between the adoption of a program and the realization of student gains. Change doesn't just happen because we want it to or expect it to. What happens in the chasm spanned by the "giant leap" is *implementation.*

Researchers have identified strategies that can help schools and school leaders successfully maneuver this "giant leap" and operate in a new way that results in improved student achievement (Hord, 1992). These strategies for successful implementation can be organized into six categories:

1. Creating an atmosphere and context for change
2. Developing and communicating a shared vision
3. Planning and providing resources
4. Investing in professional development
5. Checking progress
6. Continuing to give assistance

In this article, each of these categories is examined, with suggestions for what effective leaders do to encourage and nurture the implementation of any newly adopted program. Examples from the field are provided including individuals and groups that represent leadership in districts and schools. In these examples, leaders include superintendents, principals, teachers, central office staff, and leadership teams.

Creating an Atmosphere and Context for Change

When you think about it, school improvement or school reform really means that the school and its staff are *changing* how they do business. Changing how to do business is dependent upon the staff *learning* how to do things differently. Fundamental change flourishes in an environment where everyone is committed to learning—learning for staff first, followed by learning for students whose learning experiences are of increased intellectual quality provided by more effective teachers and administrators. In such an environment change is promoted and risk-taking is encouraged.

Teachers may feel unsure of themselves when they begin to operate in different ways. Leaders in these schools pay attention to the concerns of teachers and are never too busy to listen and interact with staff: Do teachers need more information? Assistance with preparation of lessons or materials? Assistance with using new instructional strategies?

In a school creating a context conducive to change one will see the entire professional staff (teachers and administrators) coming together to reflect on how they are working to achieve goals for students. They consider what is working well and what is not working so well, and use this information to assess how effective they have been. From this reflection and assessment, they determine where the "soft spots" are that need attention and decide the learning that they need to do to "shore up" these areas of need.

Effective leaders demonstrate their support by taking an active role in these professional learning activities alongside teachers and by encouraging teachers to take leadership roles in the identification and implementation of new initiatives. With teachers, they share power, authority, and decision making so that the entire school staff acts as a community of leaders and learners.

A superintendent that we know interacts directly and personally with his campus principals. They make plans and set goals on which the principal will work. The superintendent makes clear that risk-taking is in order so that innovation and new practice can flourish. If mistakes are made, they are accepted, if learning accrues from the errors. In this context, mistakes are not seen as debilitating events, but as part of the process of changing the school's structures and practices that will benefit students.

Developing and Communicating a Shared Vision for Change

Implementation of a new program is more effective when it is focused by a clear vision for change. Having a clear vision is more than having lofty goals. It refers to mental pictures of what a school or a classroom might look like in a changed and improved state—a preferred image of the future. Effective leaders work together with staff to create this clear image of what a new program will look like when it is functioning in the school or classrooms in a high quality way. When a staff has collaborated on developing a shared vision, each individual (teacher, principal, and paraprofessional staff alike) has a clear picture of what they must do to accomplish the desired change, but they also have a picture of how their role supports the overall program.

A compelling vision reflects the values held by the staff and stakeholders and it drives all decisions made at a school. For example, if the vision of the new instructional practice indicates that students will develop mathematical concepts by manipulating objects in math, then funds will be set aside for the purchase of manipulatives; teachers will change practice by incorporating manipulatives into lessons; and students will be seen using manipulatives in math classes.

In order to foster a common understanding about the vision, effective leaders use every opportunity to refer to the vision of the school during school and community meetings, personal interactions, and written communications.

One of the effective principals that we have studied convenes the entire school community (school staff, students, parents, business representatives) to share with them the vision of change toward which the school is working. As the constituents work on creating mental images of the change when it is in place, they gain a sense of where the school is heading and what they can do to support the efforts, not only in the classroom, but in the home and community. This community-wide shared vision increases the potential for the change to be well implemented so that its benefits can be realized in students.

Planning and Providing Resources

A vision is of no use without a clear roadmap to guide the school in realizing that vision; implementation plans give administrators and teachers clear, specific and orderly instructions to follow as they implement their initiative. Effective leaders engage their staff in the planning process and in a dialogue about the best use of people, time, and dollars to support a new program.

Planning should be less of a blueprint and more of an evolutionary process. That is not to say that a school's improvement plan is abstract, obtuse, and non-specific; but effective leaders work with teachers and adapt plans based on direct experience with what is working in moving toward the vision and what isn't. The plan evolves through interactions with participants (administrators, teachers, parents, and students) and is clearly focused on improved student performance.

How resources (dollars, personnel, time, and energy) are allocated at a school is a reflection of a school's goals and priorities as reflected in their plan. In effective schools, resources are allocated, and reallocated, in ways that maximize teacher learning, organizational learning and, thus, student learning.

An effective high school leadership team that we've worked with included the principal, teachers, parents, and a central office staff member. They took the time necessary to conduct a complete audit of the current state of affairs at the school. They researched student achievement trends, graduation trends, and course loads, and they analyzed the budget. They learned about obstacles and expectations from staff, parents, students, and community members. The team identified the key issues facing the school, combed the research for approaches that would address those issues and worked with the entire faculty to design a new way of operating based on student needs. Some former programs were dropped, funds were reallocated and staff was realigned to match the new program where all efforts were focused on having every student graduate.

Investing in Professional Development

Change efforts most often require the acquisition of new content knowledge and/or additional instructional techniques and strategies. Teachers need a way to acquire the necessary skills and knowledge, receive feedback and support as they begin to use the skills, and have opportunities to reflect on how their behaviors impact student achievement. Effective professional development affords teachers opportunities to learn together and discuss new ideas with their colleagues and review together how samples of student work have changed as a result of the new program. Teachers need risk-free opportunities to practice, view model lessons and receive coaching and feedback on the use of the new ideas or strategies.

Effective leaders use multiple forms of data such as student achievement data, survey and interview results and program evaluation data to identify needs of the staff for training and development. They encourage a deep understanding of new ideas through collegial learning opportunities rather than mechanical implementation of a program. They also participate directly in staff development sessions and take part in planning, conducting, implementing, and evaluating the effort.

Schools that place an importance on professional learning provide adequate time for staff development and follow-up. In these schools, teachers are provided support and materials they need as they implement new instructional strategies.

Enlightened districts, and there are quite a few, are providing schedules where student early release days permits the professional staff to come together regularly and frequently for "faculty study." At these times, the learning and growth of the faculty has been defined by the issues of need of the particular school. In these professional learning communities, the staff is continuously studying itself to make decisions about their needs for professional development so that their improvement is continuous and directed toward the needs of the students.

Checking Progress

Checking the progress of implementation is important so that principals and teachers can learn from their successes and setbacks. They must know what's working (and why) as well as what's creating problems in order to provide support for the change effort where and when it's needed and deal with problems appropriately. Checking progress is a leader's continual effort to "touch base" with implementers, seek input about their needs, and assess implementation progress.

School improvement efforts, no matter how well planned, will always encounter challenges and problems at some stages—some slight, some that are possibly severe. Effective schools have many ways to measure the progress they are making toward incorporating new ideas into everyday practice and to check how much they have learned. They set specific benchmarks and check them against the "preferred image for the future," the staff's vision of the program when it's being implemented fully.

Effective leaders continually check current performance against the vision. They acknowledge problems and confront them rapidly but they also recognize the positive steps being made toward a new way of operating. They develop an atmosphere of trust so that frequent visits to classrooms are a comfortable routine and are non-threatening. They gather information formally, i.e. by surveys and testing, or informally, i.e. by interactions with teachers in hallways and brief classroom visits. Feedback is frequent and constructive and it is used to make adjustments as the implementation unfolds.

When teachers develop a system of "peers helping peers," they touch base as partners with each other, serving as "critical friends" or "friendly critics." They observe each other's performance and give constructive feedback. In addition, they study students' work across classrooms and deliver helpful critiques. This interaction among staff contributes to individuals' effectiveness as well as that of the organization. In this way, everyone is involved in checking the progress of the implementation effort.

Continuing to Give Assistance

A new idea is like a seedling—it needs care and attention to flourish and mature. As teachers and administrators begin to change strategies and methodologies, there are personal and management needs that arise and continuing assistance is required. These needs must be addressed and they change as implementers move from novice to expert in their improvement efforts; assistance is structured to focus on these needs.

Effective schools provide many ways for all staff members to provide support for each other including peer coaching and mentoring. They take the information generated as they check progress and work out ways to help those who need additional assistance with incorporating new ideas into daily practice.

A leader's actions should focus on promoting implementation through coaching, problem-solving and technical assistance to individual users. Assistance can be as simple as having materials for teachers in classrooms as they are needed or as challenging as creating organizational structures that allow teachers to plan together and observe one another. Giving consistent attention and acting on problems involves enormous persistence and tenacity, and good leaders address problems from every possible angle over time.

Effective leaders also look for positive progress and directly and sincerely recognize and praise teachers. An effective source of support that is often overlooked is bringing staff together regularly to celebrate successes that teachers are experiencing as they implement the new program.

Assistance is coupled with checking progress; they go hand in glove. Central office staff can support assistance efforts by providing additional materials and the human dimension of recognizing and celebrating a school's efforts and successes along the process of implementation. Central office staff know and understand that the implementation dip suggests that before things get better, they will be challenging and difficult. Their attention to the school and their support in assistance giving is a significant boon to a school in the process of doing the hard work of changing their curriculum, instruction, assessment, or other factors.

Sustainability

If implementation proceeds well, and staff learn how to use new practices and processes in a high quality way, there is a good chance that such practices may be sustained and become routinized into the daily life of the school or district. Too often staff members learn only to use new practices in a superficial way and without deep meaning and understanding of how to adapt them for the benefit of individual students. Without deep understanding they become mechanical users of new programs and practices. The goal is not to maintain such dogmatic use, but to provide staff sufficient time and learning opportunities so that they understand the philosophical approach and rationale for the new practices that they are implementing.

Then the goal is to make certain that the implemented program has a line in the budget, new staff are given training and development activities, and resources are appropriately allocated so that the program is soundly supported. Above all, leadership in the school gives continuous attention and directs the attention of others to the value of the now not-so-new program. That the program and its practices are given support and also celebration are keys to its robust sustainability, and its capacity to contribute richly to the learning success of students.

References

Fullan, M. G. (2001). *The new meaning of educational change* (3rd ed.). New York, NY: Teachers College Press.

Hord, S. M. (1992). *Facilitative leadership: The imperative for change.* Austin, TX: Southwest Educational Development Laboratory.

Published Citation

Tobia, E. F., & Hord, S. M. (2002, March). Making the Leap: Leadership, learning, and successful program implementation. *Instructional Leader.* Austin: Texas Elementary Principals and Supervisors Association.

HANDOUT 2.1

Assessing the Degree of Implementation of the Six Strategies

Name of the Change Effort:				
Strategy	*High Degree*	*Some Implementation*	*Poor Implementation*	*None, No Implementation*
Articulates a shared vision of the change				
Creates a plan and identifies resources to achieve the vision				
Invests in professional learning about what the change is and how to use it				
Assesses the degree to which the vision of the change is implemented				
Provides one-to-one or small-group assistance to support implementation				
Creates a context that supports and encourages the change				

LEARNING MAP 2.2

Planning Strategies for a Change Effort

Outcome

Learners will create initial plans for a change effort, focused on the strategies, and explain how they will be used to cross the implementation bridge.

Assumption

A plan is not set in concrete, but it is the means by which to launch a change effort and to look long term to identify actions that will be required over the course of the change project, recognizing that adjustments will undoubtedly be needed.

Suggested Time

60 minutes

Materials

1 copy for each participant of
- Handout 2.2.a, Six Strategies for Change: Talking Points and Questions
- Ask the participants to bring the "Making the Leap" paper from Learning Map 2.1.

2 copies for each participant of
- Handout 2.2.b, Constructing a Skeletal Plan for Crossing the Implementation Bridge

Engaging in Learning

Review briefly the major points from the material of Learning Map 2.1. Ask for questions; turn the questions back to the other participants, requesting them to respond. If responses are incorrect, or none are offered, provide answers to the questions.

1. Guide the whole group in collaboratively constructing a skeletal plan for "crossing the implementation bridge." A very useful tool for considering the creation of such a map has been provided by a good colleague, Edward Tobia, of SEDL in Austin, Texas. Study this work on Handout 2.2.a, Six Strategies for Change: Talking Points and Questions. Review Handout 2.2.a independently, underlining significant points made about each strategy. After the independent review, invite table team members to circle the table and the list of strategies, giving each individual the task of leading a 2-minute interaction with the group about a strategy, focusing on points the team members considered important. Maintain a list of any questions that are generated as your table team reviews Handout 2.2.a.

   Provide time for this review and discussion, then reconvene the whole group and solicit questions they have recorded. Invite responses to questions from the whole group, and refer to Handout 2.2a when applicable.

2. Remind the participants to keep in mind the ideas we have just reviewed, for now we will turn attention to creating a skeletal map, and we will do this collaboratively with the whole group involved together. Use the scenario below to guide this activity.

   Scenario: Anywhere Middle School has 575 students in Grades 6–8, with a teaching faculty of 30. After careful study and exploration, the faculty has decided to adopt a new classroom practice—a new questioning approach that will increase their teaching proficiency and students' understanding and learning results. It is a strategy that will be used by all teachers at all levels.

3. Use Handout 2.2.b and guide participants in creating and recording a plan for identifying strategies for a 2-year implementation of the new questioning practice. For each strategy noted, an explanation should be provided for its identification. Ask participants to determine the sequence of the strategies and how long it might take to implement each specific strategy in a 2-year period of time.

   Discussion across the whole group should be entertained, with all participants invited to contribute with rationales for their ideas.

   Ask participants which of the six strategies might be the initial one, articulating a shared vision of the change or creating a context that supports and encourages the change: "Yes, let us start with a vision of what the change will look like in classrooms after it has been implemented well. . . . We start with this strategy because? Yes, we must know where we are going, or what the end point will be before we start."

   We might start with consideration of creating a context for the change and that is a good idea. Such a context requires much time and should be considered for attention at the beginning, while at the same time working on a written vision for the change; knowing where the school staff and students are going with this innovation contributes a great deal to having a context conducive to the change. Guide the group similarly through the steps, soliciting appropriate explanation for each step, referring back to the "Making the Leap" paper and to Handout 2.2.a for ideas.

   The Creating a Context strategy is one that should be considered early and actions taken to increase positive relationships, trust, respect for each other (staff), and certainly starting with being sure that the staff know each other and have opportunities to interact across their grade and content assignments. Starting with a hot fudge sundae party is a splendid idea!

   Maintaining consistent focus on the context across all the rest of the strategies is wise. After the consideration of the vision and the context, the strategies as they are listed are reasonably sequential, for one cannot plan without having a vision (describing the target picture to which the implementation will be heading) to guide the plan, and without the vision it is not possible to identify resources.

Professional learning, as we emphasize throughout this book, is the basis for change and should be a continuous process: learning what the new "way" is and how to use it, in learning sessions that are long enough to reach a learning outcome, followed by its implementation, using a cycle of such sessions over time, with opportunities to implement each before the next session.

Checking progress (and we use this term rather than *monitoring* or *assessing*, as *checking progress* is a more positive term and less threatening to implementers) is conducted to ascertain who needs what in order to support the implementers with continued assistance, the last strategy. The combination of checking progress and giving assistance is what coaches do. They learn how implementation is going with individuals and then provide them with the appropriate support needed.

These comments provide a brief introduction and can be used to guide and support the creation of an implementation plan using the research-based strategies.

When the plan is complete and questions and discussion terminated, move on to Steps 4 and 5.

4. Organize participants in small groups of three to four, whose task is to decide on an innovation (new program, process, practice—a change that is being executed in one of their schools). Use the second copy of Handout 2.2.b to provide this information and descriptors for the school, and create a plan (similar to the whole group's practice) for crossing the implementation bridge. This task provides for more independent practice, subsequent to the guided practice above. Timelines and explanations for the decisions about the plan should be noted on the handout. Allow 30–40 minutes for this activity.

5. Invite groups to present their plans to another group. After this 10-minute activity, solicit one group to share their plan with the large group; invite critiques of their plan—any critique must be coupled with an explanation or rationale. Solicit questions or needs for clarification.

   Celebrate the products.

Before dismissing, ask participants to bring their materials from sessions 2.1 and 2.2 to their next meeting.

HANDOUT 2.2.A

Six Strategies for Change

Talking Points and Questions

Strategy 1: Creating a Shared Vision of the Change

A shared mental image of the future as a result of successful implementation of the innovation

Questions:

- Ideally, what do we want our innovation to look like once it has been fully implemented?
- How do we ensure that the vision is a shared vision?
- What are some effective strategies for communicating the vision to others?

Strategy 2: Planning and Identifying Resources Necessary for the Change

The roadmap for change, and the time, tools, and staff needed to implement the innovation

Questions:

- What new roles need to be created (or existing ones realigned)?
- What time, tools, and staff will be needed for ongoing planning, professional development, and collaboration?
- How do we ensure that our plan remains up to date?
- How will we know that our plan has been implemented and is having the desired impact?

Strategy 3: Investing in Professional Development/Professional Learning

Provides implementers with what they need to know and be able to do, and provides evidence of implementation and impact

Questions:

- What professional learning does the staff need?
- How do we design and provide professional learning to meet the needs of staff throughout the process of implementation?
- What professional learning do our leaders and facilitators need to support implementation?
- How do we know that the professional learning was effective?

Strategy 4: Checking or Assessing Progress

Provides strategies to identify emerging needs, clarify questions, solve small problems, and provides evidence of implementation and impact

Questions:

- What types of data do we need?
  - Evidence of implementation
  - Evidence of impact
- How frequently should we collect implementation data?
- How do we effectively interpret the data we have/want?
- How do we communicate what we learn from the data? To whom?

Strategy 5: Providing Continuous Assistance

Strategic and targeted responses to support implementation based on identified needs

Questions:

- What forms of assistance will maintain the momentum of implementation?
- How do we sustain and improve implementation in the face of changes and challenges?
- How can we incorporate what we learn from monitoring to improve the process?
- What are possible ways to celebrate and acknowledge successes?

Strategy 6: Creating a Context Conducive to Change

Creating and nurturing a culture and climate in the organization to support implementation

Questions:

- How do we create a sense of urgency about the need for implementation of the innovation?
- How do we build a sense of mutual responsibility and accountability for implementation?
- How might structures in the school (use of time and space) be redesigned to promote a context for change?
- How might we and others model behaviors and norms that support implementation?

Source: Hord, S. M. (1992). *Facilitative leadership: The imperative for change* (p. 31). Austin, TX: Southwest Educational Development Laboratory (SEDL). Copyright 1992 by SEDL. Adapted by Ed Tobia with permission of SEDL. Reprinted by Shirley Hord with permission of SEDL.

HANDOUT 2.2.B

Constructing a Skeletal Plan for Crossing the Implementation Bridge

Name of the school:	
Grades of the school:	
Number of students/teachers:	
Name of the adopted (new) practice:	
Goal of the new practice:	
Step 1 Strategy(s)	Explanation
Step 2 Strategy	Explanation
Step 3 Strategy	Explanation
Step 4 Strategy	Explanation
Step 5 Strategy	Explanation

LEARNING MAP 2.3

Reviewing the Literature on Structural and Relational Conditions for Change

Outcome

Learners will briefly describe a selected set of contextual factors, accessed from the literature, that are valued for successfully introducing changes in organizations (schools and districts).

Assumption

In addition to their own experiences, wise leaders refer to the research knowledge base for obtaining guidance in establishing the conditions most conducive for supporting their change process.

Suggested Time

60 minutes

Materials

Chart paper, self-sticking type

Markers

Large-size sticky note pads, one for each two to three participants

1 copy of the literature on conditions for successful change (excerpted by James Roussin), for each participant:

Research briefs (4) abstracted from Hoy, Tarter, and Woolfolk Hoy (2006)

Engaging in Learning

1. Set the stage by sharing with participants that we will dig a bit deeper into the "context for successful change" strategy. Provide two quick descriptions (from your own experience) of two contrasting conditions that participants might encounter in initially entering their learning location where they will study their newly adopted program or practices, such as
   a. a dark, uncomfortably cold and undecorated room, except for the "rules for the room" posted in the middle of the major wall, or
   b. a large, airy, brightly lighted room with a bouquet of flowers on the podium and an offering of a few "sweets" at each person's seat.

   Ask participants to chat for 2 minutes about these two settings, determining which setting will be most conducive to their comfort, interest, and attention to learning. The physical conditions, of course, do not solely constitute the elements in the setting that will impact participants.

2. Guide the participants in finding their Research Briefs (1–4) and in identifying a foursome of learning partners. Note that these four papers do not constitute the entire studies of this topic; they are a noteworthy sampling. While they do not speak directly of school change, they do describe conditions or factors in schools that influence staff to consistently seek change and improvement in order to develop and/or maintain quality teaching and learning. Before starting this learning activity, create four charts with the following headings and post them in the room:

Academic Emphasis of Schools	Collective Efficacy	Faculty Trust in Parents and Students	Academic Optimism

   Request one person in each group of four to serve as facilitator for the first research brief. Each individual reads the same brief, marking parts if they wish, and then engages in the activity of *say something.* In this activity, there is no cross-talk—each person (supported by the facilitator) relates his or her reactions, thoughts, or feelings resulting from reading the research brief. For closure, the facilitator summarizes crisply the key ideas by the group, records them on a large sticky note, and places it on one of the four charts on the wall, designated for that research brief. The remaining briefs are reviewed in the same way with a different individual facilitating the four-person groups.

3. Focus the entire group of participants on each of the large charts of sticky notes. Discuss the ideas, and derive a summary of each brief's concepts or ideas. A recorder should indicate these ideas on the large charts.

4. Show how these "Big Ideas" can be used in a change effort by inviting participants to arrange in pairs to reflect and brainstorm applications to their work. The applications should be jotted on sticky notes and posted on a blank chart near each of the research briefs posted on the wall.

5. Provide closure by collecting the two sets of the four charts, promising to reproduce them and send them to all participants. Ask participants to engage in further reflection with colleagues at their school or district, generating applications of the ideas from the four research briefs.

Future Application

Ask participants to keep in their learning journal the reproduced charts sent to them. These charts will be used to collect a running "tab" of conditions, as they are identified across many of the lessons, and will be given additional focus and attention at a future date. The use of CBAM concepts and tools adds significantly to the context that supports and encourages change—thus, keep this material handy for additions.

RESEARCH BRIEF

Academic Emphasis of Schools *(Paper 1)*

Academic emphasis is the extent to which the school is driven by a quest for academic excellence—a press for academic achievement. High but achievable academic goals are set for students, the learning environment is orderly and serious, students are motivated to work hard, and students respect academic achievement (Hoy & Miskel, 2005; Hoy, Tarter, & Kottkamp, 1991).

Hoy and his colleagues (1991) were first to demonstrate that the collective property of academic emphasis of the school was positively and directly related to student achievement in high schools while controlling for SES (socioeconomic status). Whether school effectiveness was conceived as the commitment of teachers to the school, the teachers' judgment of the effectiveness of the school, or actual student test scores, academic emphasis remained a potent force. At both middle school and high school, academic emphasis and achievement were positively related, even after controlling for socioeconomic factors (Hoy, Tarter, & Bliss, 1990; Hoy & Hannum, 1997; Hoy and Sabo, 1998).

The findings are the same for elementary schools. Goddard, Sweetland, and Hoy (2000), controlling for SES, school size, student race, and gender, used hierarchical linear modeling to find academic emphasis an important element in explaining achievement in both math and reading. The authors concluded, "Elementary schools with strong academic emphases positively affect achievement for poor and minority students" (p. 698).

Alig-Mielcarek and Hoy (2005) considered the influence of the instructional leadership of the principal and the academic press of the school. They also found that academic emphasis was significant in explaining student achievement, even controlling for SES. They found that academic emphasis of the school, not instructional leadership, was the critical variable explaining achievement. In fact, instructional leadership worked indirectly, not directly, through academic press to influence student achievement.

The results are consistent, whether the level was elementary, middle, or secondary: academic emphasis is a key variable in explaining student achievement, even controlling for socioeconomic status, previous achievement, and other demographic variables.

Academic Emphasis

The one goal that virtually everyone shares for schools is academic achievement of students. The reform and accountability movements have promoted a press toward the academic achievement of all students (No Child Left Behind). The focus of schooling is clear: it is academic. A push for academic achievement, however, in an environment where teachers do not feel efficacious, is a recipe for frustration and stress. The challenge is to create school conditions in which teachers believe they and their students are up to the task. How might this be done? Principals move a school by example. They celebrate the achievements of students and faculty, especially the academic ones. An emphasis on the honor roll, national honor societies, and exemplary student work of all kinds are examples of behaviors that foster academics. To be sure, this is an old list, but in conjunction with building efficacy and trust, these activities take on new strength.

Source: Hoy, W. K., Tarter, C. J., & Woolfolk Hoy, A. (2006). Academic optimism of schools: A force for student achievement. *American Educational Research Journal, 43,* 425–446. Used with permission.

RESEARCH BRIEF

Collective Efficacy *(Paper 2)*

Social cognitive theory (Bandura, 1977, 1997) is a general framework for understanding human learning and motivation. Self-efficacy, a critical component of the theory, is an individual's belief in her or his capacity to organize and execute the actions required to produce a given level of attainment (Bandura, 1997). Efficacy beliefs are central mechanisms in human agency, the intentional pursuit of a course of action. Individuals and groups are unlikely to initiate action without a positive sense of efficacy. The strength of efficacy beliefs affects the choices individuals and schools make about their future plans and actions.

Student achievement and sense of efficacy are related. Researchers have found positive associations between student achievement and three kinds of efficacy beliefs: self-efficacy beliefs of students (Pajares, 1994, 1997), self-efficacy beliefs of teachers (Tschannen-Moran, Woolfolk Hoy, & Hoy, 1998), and teachers' collective efficacy beliefs about the school (Goddard, Hoy, & Woolfolk Hoy, 2000). We focus on collective efficacy of schools and student achievement because collective efficacy is a school property amenable to change.

Within schools, perceived collective efficacy represents the judgments of the group about the performance capability of the social system as a whole (Bandura, 1997). Teachers have efficacy beliefs about themselves as well as about the entire faculty. Simply put, perceived collective efficacy is the judgment of the teachers that the faculty as a whole can organize and execute actions required to have a positive effect on students (Goddard, Hoy, et al., 2004).

Bandura (1993) was first to show the relationship between sense of collective efficacy and academic school performance, a relationship that existed in spite of low socioeconomic status. Schools in which the faculty had a strong sense of collective efficacy flourished, whereas those in which faculty had serious doubts about their collective efficacy withered—that is, declined or showed little academic progress. Continuing research has provided support for the importance of collective efficacy in explaining student achievement. Goddard, Hoy, et al. (2000) supported the role of collective efficacy in promoting school achievement in urban elementary schools. They hypothesized that perceived collective efficacy would enhance student achievement in mathematics and reading. After controlling for SES, they found that collective efficacy was significantly related to student achievement in urban elementary schools.

Hoy, Sweetland, and Smith (2002), continuing this line of inquiry and using collective efficacy as the central variable, predicted school achievement in high schools. They found collective efficacy was the key variable in explaining student achievement. They found, in fact, that it was more important than either socioeconomic status or academic press. Hoy and his colleagues (2002) concluded, "School norms that support academic achievement and collective efficacy are particularly important in motivating teachers and students to achieve. . . . [H]owever, academic press is most potent when collective efficacy is strong" (p. 89). That is, academic press works through collective efficacy. They further theorized that when collective efficacy was strong, an emphasis on academic pursuits directed teacher behaviors, helped them persist, and reinforced social norms of collective efficacy.

In a similar vein, Goddard, LoGerfo, and Hoy (2004) tested a more comprehensive model of perceived collective efficacy and student achievement. They learned that collective efficacy explained student achievement in reading, writing, and social studies, regardless of minority student enrollment, urbanicity, SES, school size, and earlier achievement. Research has consistently demonstrated the power of positive efficacy judgments in human learning, motivation, and achievement in such diverse areas as dieting, smoking cessation, sports performance, political participation, and academic achievement (Bandura, 1997; Goddard, Hoy, et al., 2004).

Collective Efficacy

Collective efficacy is grounded in Bandura's (1997) social cognitive theory; hence, we turn to his sources of efficacy for ideas about how to build collective efficacy in schools. The sources of self-efficacy are mastery experiences, vicarious experiences, social persuasion, and affective states, each of which conveys information that influences teacher perceptions about the school (Bandura, 1993, 1997; Goddard, Hoy, et al., 2004; Pajares, 1997). For example, let's consider a school with a poor graduation rate. A neighboring district has implemented a successful program for at-risk students. The principal is in the position to orchestrate the transfer of the neighbor's success to his or her school. In so doing, the school is engaged in a self-regulatory process informed by the vicarious learning of its members and, perhaps, the social persuasion of leaders. Modeling success and persuading teachers to believe in themselves and their capabilities is a reasonable route to improve collective efficacy and enhance academic optimism (Bandura, 1997; Goddard, Hoy, et al., 2004).

Source: Hoy, W. K., Tarter, C. J., & Woolfolk Hoy, A. (2006). Academic optimism of schools: A force for student achievement. *American Educational Research Journal, 43,* 425–446. Used with permission.

RESEARCH BRIEF

Faculty Trust in Parents and Students *(Paper 3)*

Faculty trust in parents and students is the third school property that is related to student achievement. Faculty trust in parents and students is a collective school property in the same fashion as collective efficacy and academic emphasis. Surprisingly, trust in parents and trust in students is a unitary concept. Although one might think that the two are separate concepts, several factor analyses have demonstrated they are not (Hoy & Tschannen-Moran, 1999; Goddard, Tschannen-Moran, & Hoy, 2001). Furthermore, Bryk and Schneider (2002) make the theoretical argument that teacher–student trust in elementary schools operates primarily through teacher–parent trust.

Trust is one's vulnerability to another in the belief that the other will act in one's best interests. Tschannen-Moran and Hoy (2000), after an extensive review of the literature, concluded that trust is a general concept with at least five facets: benevolence, reliability, competence, honesty, and openness. Although it is theoretically possible that these facets of trust may not vary together, the research on schools shows all five facets of trust in schools do indeed vary together to form an integrated construct of faculty trust in schools, whether the schools are elementary (Hoy & Tschannen-Moran, 1999; Hoy & Tschannen-Moran, 2003) or secondary (Smith, Hoy, & Sweetland, 2001). Thus, we defined faculty trust as the group's willingness to be vulnerable to another party based on the confidence that the latter party is benevolent, reliable, competent, honest, and open (Hoy & Tschannen-Moran, 2003).

Cooperation and trust should set the stage for effective student learning, but only a few studies have examined this relationship. Goddard et al. (2001) examined the role of faculty trust in promoting school achievement of urban elementary schools. Using a multilevel model, they demonstrated a significant direct relationship between faculty trust in students and parents and higher student achievement, even controlling for socioeconomic status. Like collective efficacy, faculty trust was a key property that enabled schools to overcome some of the disadvantages of low SES.

Hoy (2002) examined the trust-achievement hypothesis in high schools and again found that faculty trust in parents and students was positively related to student achievement while controlling for SES. He theorized that trusting others is a fundamental aspect of human learning because learning is typically a cooperative process, and distrust makes cooperation virtually impossible. When students, teachers, and parents have common learning goals, then trust and cooperation are likely ingredients that improve teaching and learning.

Finally, Bryk and Schneider (2002) performed a three-year longitudinal study in 12 Chicago elementary schools. Using HLM models, survey and achievement data, and in-depth interviews, they concluded relational trust was a prime resource for school improvement. Trust and cooperation among students, teachers, and parents influenced regular student attendance, persistent learning, and faculty experimentation with new practices. In brief, trust among teachers, parents, and students produced schools that showed marked gains in student learning, whereas schools with weak trust relationships saw virtually no improvement. The research of Bryk and Schneider, and that of Hoy and his colleagues (2006), reinforce each other in the common conclusion that faculty trust of students and parents enhances student achievement.

Trust in Parents and Students

There is some research on family and community involvement in schools (cf., Epstein, 1989). There is little systematic research on how to build authentic trust, however. Faculty trust in students and parents can be promoted through useful interchanges, both formal and informal, between parents and teachers. Making the most of vicarious learning, for example, a school can respond to a lack of trust and community participation in school activities by emulating the practices and procedures of magnet schools, which are known for their parental cooperation and involvement. But much more research is needed about what programs and factors support the development of teachers' trust in parents and students.

Such examples demonstrate how changes in social perceptions influence what actions organizations choose to pursue. Collective perceptions about efficacy, academic emphasis, and trust shape the school's normative environment and can be developed through experiences that convey their value.

Source: Hoy, W. K., Tarter, C. J., & Woolfolk Hoy, A. (2006). Academic optimism of schools: A force for student achievement. *American Educational Research Journal, 43*, 425–446. Used with permission.

RESEARCH BRIEF

Academic Optimism *(Paper 4)*

Three collective properties—academic emphasis, efficacy, and trust—are not only similar in their nature and function, but also in their potent and positive influence on student achievement. The three concepts have much in common. In fact, Hoy and his colleagues (Hoy, Tarter, & Woolfolk Hoy, 2006) demonstrated that the three collective properties worked together in a unified fashion to create a positive academic environment called academic optimism.

Many conceptions treat optimism as a cognitive characteristic: a goal or expectancy (Peterson, 2000; Snyder et al., 2002). Our conception of academic optimism includes both cognitive and affective dimensions, and adds a behavioral element. Collective efficacy is a group belief or expectation: it is cognitive. Faculty trust in parents and teachers is an affective response. Academic emphasis is the press for particular behaviors in the school workplace (Hoy et al., 2006). Hoy and his colleagues concluded, "Collective efficacy reflects the thoughts and beliefs of the group; faculty trust adds an affective dimension, and academic emphasis captures the behavioral enactment of efficacy and trust" (p. 14). Academic optimism is a rich picture of human agency that explains collective behavior in terms of cognitive, affective, and behavioral dimensions.

When the faculty believes it has the capability to organize and execute actions for a positive effect on student achievement, it emphasizes academic achievement, and academic emphasis in turn reinforces a strong sense of collective efficacy. In sum, all the elements of academic optimism are in transactional relationships with each other and interact to create a culture of academic optimism in the school.

Hoy and his colleagues (2006) chose the term academic optimism to reflect beliefs about control in schools. They explain as follows:

> Optimism is an appropriate overarching construct to unite efficacy, trust, and academic emphasis because each concept contains a sense of the possible. Efficacy is the belief that the faculty can make a positive difference in student learning; teachers believe in themselves. Faculty trust in students and parents is the belief that teachers, parents, and students can cooperate to improve learning, that is, the faculty believes in its students. Academic emphasis is the enacted behavior prompted by these beliefs, that is, the focus is student success. Thus, a school with high academic optimism is a collectivity in which the faculty believes that it can make a difference, that students can learn, and academic performance can be achieved. (Hoy et al., 2005)

Optimism

The research on individual optimism suggests some ideas about encouraging a culture of optimism in schools. Peterson (2000) found that optimism is thwarted by stress, so decreasing stress should support optimism. Teachers can lower their stress by increasing their agency and control through appropriate participation in decisions that affect their school lives (Hoy & Tarter, 2004).

People learn from models because the observation of successful performance in others promotes an acquisition of their beliefs and the actions. The most effective models are those who seem competent, powerful, prestigious, and similar to the observer (Pintrich & Schunk, 2002). Vicarious and observational learning are sources of optimism. Thus teachers can serve as models for each other. The way school problems are discussed should convey the possibilities for resolution rather than defeatism. Novice teachers, for example, should hear optimistic approaches to teaching rather than a sense of passive helplessness in teachers' lounges and school hallways.

Our analysis is a promising clarification of the linkages within schools that influence student achievement. Although our data are drawn from high schools, we believe the findings are applicable to elementary and middle schools because the three elements of academic optimism have explained learning in those settings as well. Clearly, more research in a variety of school settings is necessary to build a comprehensive theory of academic optimism of schools.

Academic optimism is especially attractive because it emphasizes the potential of schools to overcome the power of socioeconomic factors to impair student achievement. There is a real value in focusing on potential with its strength and resilience rather than pathology with its weakness and helplessness. Optimism attempts to explain and nurture what is best in schools to facilitate student learning. This simple conclusion should encourage teachers and principals to move forward with confidence.

Source: Hoy, W. K., Tarter, C. J., & Woolfolk Hoy, A. (2006). Academic optimism of schools: A force for student achievement. *American Educational Research Journal, 43,* 425–446. Used with permission.

LEARNING MAP 2.4

Assessing Change Readiness

Outcome

Learners will describe five change readiness dimensions for determining staff willingness and capacity to participate in implementing a change.

Assumption

Change readiness is a necessary and often critical stage that must be addressed before starting any implementation.

Suggested Time

45 minutes

Materials

Chart paper, self-sticking type

Markers

One sticky dot for each person

1 copy for each participant of
Handout 2.4, Readiness for Change

Engaging in Learning

1. Set the stage by asking participants to recall a time when a change took place and staff members were not happy about it.
2. Invite participants to create a T chart in their notebook. On the left side of the T chart, identify the behaviors that staff expressed when they were not positive about the change. On the right side of the T chart, identify what early support staff may have needed in order to be more responsive to the change.

Behaviors	Early Support

3. Ask participants to find a partner and share what they identified on their T chart.

4. Ask participants to return as a whole group and share ideas and examples of what early support staff may need so they are more responsive to a change. Post responses on chart paper.

5. Give each participant one sticky dot, and ask each person to place his or her dot on one early support that may be most important in preparing staff for a change.

6. Invite participants to share what they notice about where the majority of the dots landed. Ask why early support in those areas may be important in creating a readiness for a change.

7. After the conversation, ask participants to locate Handout 2.4, Readiness for Change. This handout identifies key indicators of readiness that informs change leaders when potential implementers are ready to begin implementing a change.

   Ask participants to read and underline important factors in the readiness dimensions and indicators listed on the handout—factors that they deemed most critical or important to their change efforts in their schools.

   They might also underline examples of early support they identified on the right side of the T chart that are similar to those listed in Handout 2.4.

   When they have finished reading and reviewing the handout, ask them to find a learning partner and, together, describe to each other the five dimensions of change readiness and why these dimensions are important in guiding change.

8. Invite the group to explore ideas in how the Readiness for Change handout might be used in determining staff willingness and capacity for a change.

Future Application

Make a large poster-size version of the Readiness for Change handout, and use it as a guide when entering the start of any change. You might suggest change leaders reference the chart as a checklist for assessing organizational readiness before implementation.

HANDOUT 2.4

Readiness for Change

The capacity to create readiness for change, manage the change process, implement innovations effectively, and establish reliable and enduring indicators of progress is largely missing in nearly all State education systems. Creating this infrastructure is an essential part of effectively using evidence-based programs and other innovations to benefit students.

–Fixsen, Blase, Horner, & Sugai (2009)

Readiness Dimensions	*Readiness Dimension Indicators*
Relevance and Meaning: *Make a compelling case for the change, or identify the benefits of the innovation.*	❑ There is ample data to justify the need for this change ❑ Staff have had plenty of opportunity to dialogue on the "whys" for this change ❑ This change is *not* being driven by a crisis mindset ❑ This change aligns with the organization's core values and mission ❑ There is anecdotal evidence from staff expressing why this change is important ❑ There is evidence that a culture of trust exists with staff around this change ❑ ______
Consensus and Ownership: *Engage staff so there is an owning of the desired change.*	❑ Staff express ownership for this change ❑ Staff say they are willing to commit time and energy toward this change ❑ This change was *not* driven by a top-down mandate and one-way communication ❑ Staff see this change as making a significant difference that will bring results ❑ Parents are strong supporters for this change effort ❑ There is shared responsibility and collective trust for this change ❑ ______
Scope and Culture: *Define the scope of the change and the impact it will have on the school's culture, current mindsets, and behaviors.*	❑ Advocacy for this change has been sensitive to the organization's culture ❑ Staff mentally, emotionally, and physically embrace this change ❑ Change leaders have been respectful and sensitive in helping staff make sense of this change over time ❑ This change aligns well with innovations that have been recently implemented ❑ This change will not overwhelm the current workload of staff ❑ There are examples of change leaders modeling this change ❑ ______
Structure and Coherence: *Determine change leadership roles, structure, and decision making, and how the change will interface with district operations.*	❑ The right stakeholders have participated in making the decision for this change ❑ Leadership has identified key roles to support the change in moving forward ❑ Staff understand how future decisions will be made around this change ❑ There are appropriate resources (financial, time) to implement the change ❑ This change is feasible and the right resources are in place to sustain it ❑ Frequent and adequate communication with feedback have guided this change ❑ ______
Focus, Attention, and Letting Go: *Assess where to focus attention based on data, and determine what can be let go in order to create room for the change.*	❑ Change leaders have determined what past initiatives/practices can be let go in order to make room for this change ❑ There is a reasonable time line established for this change to support its full implementation ❑ There is a clear understanding by staff of what the change is going to be ❑ Staff understand the expectations and demands for this change ❑ There are indicators established for this change to identify early successes ❑ The appropriate technology tools are available to support this change ❑ ______

3

Innovation Configurations

Creating a Vision of the Change

EVERYWHERE BEGINS CREATING A VISION OF ITS CHANGES

After Bob Yurself and his leadership team learned about the six strategies for change, they were convinced that readiness was a critical factor in creating a context for change strategy and for starting to build the implementation bridge for guiding the new math reforms. All of his team members could look back on past examples of change and recognize how often EveryWhere made the mistake of taking a giant leap toward implementation without first addressing the strategies and learning what needed to take place so implementation would be more facile. The recent middle school math adoption was a perfect example of taking a giant leap and leaving those who had to make sense of the change in a place of confusion and frustration. Bob's team could see why it was important to create an atmosphere and context for change—where all individuals expected to implement the mathematics innovations would be involved in collegial conversations for learning about and making sense of the implementation. They were ready to do this change right!

Bob's team now had the understanding they needed to assess the readiness conditions for change, and there was clear evidence that EveryWhere was ready to take the next steps in implementing the math changes. Bob and his team were excited to move forward into the next stage of implementation.

Bob went back to his garden metaphor. He thought about the process he experienced in creating his own garden and how it began as a place of wants and wishes through the process of imagining. He remembered taking out pen and paper and how he created a visual picture (with lots of "wordy" notes) of the garden he envisioned. He also recalled how often he had to use that big eraser on his desk so the garden he imagined would match the existing conditions (garden zone, soil, space, and light). By creating a visual representation (mental image) of his garden, Bob learned what he would have to do to make his vision a reality. He realized the same is true when preparing to implement a change. There had to be a vision for the change.

Bob noticed how his team's learning about the concepts and strategies of the Concerns-Based Adoption Model (CBAM) was informing their deep understanding of how to guide change and to construct and conduct the strategic work of implementation. The next step for Bob and his team was to become very clear about what the new mathematics approaches would look like when implemented in classrooms. To that end, the leadership team and the school change leaders had to learn about the Innovation Configuration map. The creation of an IC map would help EveryWhere staff become precise in understanding clearly what the new math innovations were and how to use them productively. Bob thought that the staff would be able to envision themselves (in the maps) as they were implementing the math innovations and calibrating their journey of change. Bob was motivated for his team, and for himself, to engage and to learn from the next learning maps. He knew that the sessions about IC maps would be very valuable to the team leaders because the IC has multiple applications to the change process, and these fit readily into a number of the strategies. For example:

- The IC map is a written document that reveals precisely the vision of the change, an initial strategy. Thus, the creation of a map of the innovation to be implemented is a key action to take. The strategies are the actions necessary to move across the implementation bridge, and the tools (IC, Stages of Concern, and Levels of Understanding) provide the insights, the understandings, and the means by which to design or craft, modify or adjust, the strategy as it is being used for implementation.
- The IC map is readily applied to the strategy on visioning the change, as noted earlier, and for developing the plan for the implementation of the innovations. Planning is a second strategy in which the IC map is used as a basis for constructing the steps necessary to reach full implementation of the innovation. The third strategy, investing in professional development/professional learning, also rests on the IC map, for the map informs the users of what they should be doing with the innovation and thus what they are required to learn in professional learning sessions.
- The IC map is further valuable for checking the progress of implementation. The implementer can utilize the map to identify his or her proximity to the ideal form of the innovation, and change coaches can be more informed about how to support the implementer in moving further toward ideal practice.

There is much to be learned in Chapter 3!

LEARNING MAP 3.1

Articulating the Need for Precision About the Change

Outcome

Learners will explain the imperative for creating a mental image—a written picture—of the change when it is in operation.

Assumption

Too often change efforts provide a very shabby description of the change to which the organization is moving, through implementation. Only when the change target is clearly defined can participants see clearly where they are going and have any real chance of arriving there.

Suggested Time

60–90 minutes

Materials

1 copy for each participant of

Descriptions of Teachers A, B, and C

Interviewer protocol

Handout 3.1, Analysis and Comparison of Three Classrooms' Practices

Engaging in Learning

Note: It might be helpful to invite four members of the learning group to work with you in preparing for this activity prior to its taking place. We would suggest you give them their assigned parts (Step 1) the day before the whole group meets for this activity so they can be well prepared for their roles.

1. Distribute a different description—Teacher A, B, or C—to three of the group participants. Ask the individuals to study carefully the classroom teacher who they will portray and be ready to be interviewed about New Math, playing the role of that teacher, in a 4- to 5-minute activity. Select an individual to be the interviewer, provide the interview protocol to this person, and direct him or her to prepare to use the questions with the three teachers. Inform the interviewer that there will be three rounds of interviews lasting approximately 2–5 minutes each. Ask all four group members to remain faithful to the script, not adding or deleting commentary.

2. Distribute Handout 3.1 to each of the remaining participants, who are organized in pairs. Explain that the three individuals will portray classroom teachers. The observers will listen carefully to the three interviews of the teachers, with the task of ascertaining what the major components

or parts of the New Math Program look like from the teachers' perspectives. You can jot down components you heard for each teacher in the space provided in Handout 3.1.

3. Direct the dyads to remain with their partners. Provide them with copies of the interview protocol and the descriptions of the teachers. Ask them to read the interview protocol and the teachers' descriptions, then check their handout where they have noted the components (major parts) and add notes in the cells to describe what each of the three teachers described as his or her work for the New Math.

4. After hearing the math program described by the three teachers, and reading their descriptions, name four components of the math program that should be addressed during implementation that are based on the three teachers' use. Write the name of each component in the four cells in the first column on Handout 3.1.

5. Solicit participants around the room to share information, to be written in the cells, checking for accuracy with the teachers who are playing their roles, to be sure that each teacher is described according to the role-plays.

   Teacher A's information on the chart should include the following (from the interview responses):

Component	Objectives	Uses textbook's chapter objectives in sequence of the text
	Materials	Heatherton math textbook
	Assessment	End-of-chapter test and district mastery test
	Next Steps	When the class masters the objective, moves to next chapter and its objective

   All three teachers have articulated the same four components, noted in the interview responses. Their cells should now be filled with information from the interviews and descriptions; the cells for Teachers B and C should be filled in similarly to those for Teacher A.

6. Ask the group, "What *is* the New Math Program?"

   Note: The three quite different descriptions will lead participants to realize there can be very different perspectives about what represents the new program—given the three descriptions by the teachers, who are next-door neighbors.

7. Ask the group, "What is needed to ascertain what the new program is?" The response should be a specific articulation of the new program. In our case, suggest an Innovation Configuration map that will describe and inform us of the new program or practice, and guide our implementation of new practices, programs, processes, and so on. This instrument will be the focus of our next learning.

INTERVIEWER PROTOCOL

You will interview Teacher A, Teacher B, and Teacher C separately, using the interview protocol below. Study the three teachers' descriptions so as to be familiar with them, asking the teachers for any clarifying information, so that each of them portrays the teacher as directed by their written description.

Use this protocol with each of the three teachers, pausing so that participants may hear and understand clearly the teachers' responses.

- Ask, "Are you using the New Math Program?"
- Ask, "Would you describe for me what I might see if I were observing you using the New Math Program?"
- As each teacher describes a component of the program that is being used, summarize it back to him or her in the same way that he or she described it. If the teacher strays off the topic, gently bring the teacher back to the question.
- To conclude, thank the teacher for his or her help and usher the teacher out of the interview.

TEACHER A

I really like the New Math. The Heatherton math textbook is organized into chapters, with each chapter being the topic of a learning objective.

These chapter objectives guide my teaching, and I teach them in the sequence of the textbook.

I use the material, activities, and exercises in the textbook to teach the objective. When I believe that the class has mastered the objective, I use the end-of-chapter test to verify if they know that objective, and then I move on to the next chapter and objective. Occasionally, I use the district's mastery test to discover if all the kids are doing okay, since the district likes to keep some sort of handle on the new program.

TEACHER B

Oh, yeah, I can tell you about my teaching the new program. I am pretty excited about it, and I am pretty sure it will be a really good program for my third graders when I really know what I am doing with all the stuff.

You see, there is a set of district objectives that are to guide our teaching sequence. I've been studying them, and then I look for and identify the materials that I will use to teach each objective.

We have the Heatherton textbook as a basic set of teaching materials, but there is also a supplemental kit that is to be used to teach the objectives that are not covered in the textbook. So this is a new way to organize, especially when everything is not in one textbook.

Anyway, when I think that the students have accomplished the objective that we are working on, I use the district assessment test to find out who "got it" and who didn't.

As a result of that test, and my classroom observations and students' daily classroom work, I determine if everyone is ready to move on. Inevitably there are two or three who aren't ready, so I arrange them in a small group to reteach while the rest move to the next objective.

Did I answer your question okay?

TEACHER C

Come in . . . yes, I have a few minutes to chat with you about the Math Program.

You see, I have been teaching third-grade math for 18 years, and therefore, I am pretty familiar with what these students need to know in math.

I have my own list of objectives that these students should achieve in the third grade. I use this list to tell me where to start at the beginning of the year, and which objective to teach next—in my list.

Over the years, I have made all of my materials for teaching the lessons, and I have them filed by objective in my cabinet so that I can pull them at any time. This has been very helpful; I don't spend so much time puzzling about how I will teach or what I will teach with. I think we may have a new math textbook, but I haven't worried about studying it yet, as we have a new social studies curriculum that was given to us last year, and I am still working on getting that all straightened out.

Anyway, after I have used my materials, then I test the students with the tests that I have made, I record their scores on the test, and then I move on to the next objective.

Would you like to come and visit and see how this all works? You would be very welcome.

HANDOUT 3.1

Analysis and Comparison of the Three Classrooms' Practices

After hearing the math program described by the three teachers, name four of its components that should be addressed during implementation that are based on the three teachers' use. Write the name of one component at the start of each of the four rows. For each teacher, in the appropriate cell, write what he or she is doing relative to that component.

	Teacher A	*Teacher B*	*Teacher C*
Component:			
Component:			
Component:			
Component:			
Reactions/Conclusions:			

LEARNING MAP 3.2

Identifying Structures of an Innovation Configuration Map

Outcome

Learners will identify and define the two major structures of an Innovation Configuration (IC) map.

Assumption

Implementers and users of a change in practice need a precise understanding of what the new way looks like in use in order to implement it well. An IC map provides mental images and text pictures of the change, or innovation, as it is expected to appear in practice. Understanding the two major structures of the map initiates this understanding.

Suggested Time

60–90 minutes

Materials

1 copy for each participant of
Handout 3.2, IC Map for the New Math Program

1 set of the three teacher descriptions (A, B, C) used in the previous learning session, for each three participants

Engaging in Learning

1. Organize participants in groups of three. In our previous learning, we observed that three teachers had very different ideas and expectations for what they should be doing with the New Math Program. They had next to no guidance about what or how the new program should be delivered to students. Invite participants to locate Handout 3.2.

2. We have an IC map of the New Math Program. Let's have a look at the map. Please understand that this is a very simple example of an IC map, for getting us started using this tool. Maps are typically longer and fuller, with more description. Please notice that there is a component (the big pieces of the new program) listed above each row in the handout. How many components do you see?

   Yes, there are five components. Would you please read these five as a choral response, and hold up fingers to show me how many components there are on this map. Five, thank you. Now, read those components aloud: Selects Objectives, Uses Materials, Engages Students in Learning, Assesses Progress, Identifies Next Steps.

3. Now, look at the first component, Selects Objectives. How do the descriptions in the cells across that continuum inform us?

Yes, they give us a variety of ways that the objectives are selected by the teacher. Read the remainder of descriptions along the continuum; they all describe for us the variety of ways that teachers are likely to be working with that particular component of the New Math Program.

4. On this map, we see the five *components* of the program and the *variations*, the possible ways that a teacher may be implementing each of the components, giving us a picture of the program in each classroom.

   On the back of your IC map, write a brief definition of *component* and *variation*, as structures of the map.

5. Take a short break to stretch while we distribute a set of each of the three teachers' descriptions that we used last time, and please bring your IC map for the New Math Program to use, in 5 minutes.

   During the break, distribute the sets of the three teachers' descriptions, one for each three-person group.

6. After the break, invite the three-person groups to return to their tables. Remind them that each of them has an IC map for the New Math Program, and for each group there is a set of the three teachers' explanations of how they were using the new math, which we studied last time we met. The task now is to put one teacher's identifier (A, B, or C) on each of your three IC maps.

   Using the teacher description, mark a map for that teacher, finding on the IC map where the description of the teacher's work fits on the continuum. The groups should work on these cooperatively, discussing why a particular cell is marked and the evidence from the teacher's description that dictates that marking.

7. Share findings across the whole group. One thing learned is that the interviewer did not ask about how the teacher was teaching, or how he or she was engaging students in learning. So this entire continuum should not be marked for any teacher, for there are no responses to it. (When an interviewer is in conversation with a respondent/teacher, the initial question is open-ended—tell me about your use of the new program; what would I see? If the respondent fails to report on a component, the interviewer should ask a question to solicit information about the respondent's use of this component in the classroom.)

Markings for the teachers:

Teacher A:

Component 1—Variation 2 (no reference to district objectives, uses objectives of text chapters, thus "other published documents")
Component 2—Variation 2
Component 4—Variation 2
Component 5—Variation 2

Teacher B:

Component 1—Variation 1 (does not add others; interviewer might ask a specific question about this to find out)

Component 2—Variation 1 (same as above)
Component 4—Variation 1
Component 5—Variation 1 (not sure if teacher uses new material to reteach; interviewer should ask about this)

Teacher C:

Component 1—Variation 3
Component 2—Variation 3
Component 4—Variation 3 (classroom observation is not noted; interviewer should ask about this)
Component 5—Variation 2

There is wide variation in how the teachers are using the new math, and this should be an obvious observation. The parenthetical notes about the variations (in the list above) suggest that the interviewer be well informed about the IC map in order to check out all the factors explicitly, so as to know more precisely what the teacher is doing—and thus how to support the teacher in improved use of the program.

Inform the group that the ideal, most valued, or exemplary practice is Variation 1, on the left, on each continuum. This variation for each component is the goal of the program—what the creators of the program would like to see in the classroom. This being the case, what does this map tell us, and how might we use the information that we have gained?

8. Invite summary statements from the group:

Without a precise picture of what any new way looks like in action, in the setting where it is intended, a great many variations will undoubtedly appear—as we have seen in the teachers' descriptions.

The IC map can give us a picture of how the new program should look when it is being implemented by its intended user.

It also provides variations of implementation (or possible expected pictures) as the user learns what the new way is and how to use it (nearly no one uses a new practice in the #1 ideal way at their first effort).

Learning how to do it is imperative so that the IC map is NOT an evaluation tool, but a *growth-inducing* instrument, helping the implementer know what to learn to do.

And what are your brief definitions of *component* and *variation* that you wrote on the back of your map? Solicit responses from the group, ending with definitions that approximate these:

Component: the major features of an innovation (when it is in use) that describe how it is used; components most frequently are materials usage, teacher behaviors, and student activities.

Variations: the different ways in which the components can be operationalized, such as: Materials = program materials, teacher-made materials, commercial textbook, and so on; Grouping = homogeneous grouping, heterogeneous grouping, individualization.

How do we get an IC map if we don't have one? That is the focus of our next learning session.

HANDOUT 3.2

IC Map for the New Math Program

Teacher ______________________

Component 1: Selects Objectives

(1)	(2)	(3)	(4)
Selects objectives, in sequence from the district list, and may add objectives to address the needs of particular students.	Identifies objectives from other published documents that cover the district list.	Refers to other sources for objectives not related to the district list.	

Component 2: Uses Materials

(1)	(2)	(3)	(4)
Uses Heatherton textbook, district supplemental materials, and adds other items to increase student interest and mastery.	Stays strictly within the Heatherton textbook.	Uses other materials collected from teaching experience.	Engages randomly with no systematic set of materials.

Component 3: Engages Students in Learning

(1)	(2)	(3)	(4)
Encourages students to engage in a variety of learning strategies to meet the particular objective and specific students' needs.	Leans heavily on lecture and text assignments, with students self-checking their work.	Maintains careful daily attention to the scope and sequence of the program in order to cover the materials and objectives.	

Component 4: Assesses Progress

(1)	(2)	(3)	(4)
Observes students' daily work, provides weekly tests as benchmarks, and uses district assessments for final evidence of mastery.	Uses the Heatherton text's end-of-chapter tests routinely, and occasionally employs the district mastery test.	Relies on classroom observation of students' work and on teacher-constructed tests.	Employs no regular or systematic assessments.

Component 5: Identifies Next Steps

(1)	(2)	(3)	(4)
Moves students who have mastered current objective to the next objective, and reteaches—using new material—those who have not mastered.	Moves all students along to the next objective in order to cover the program and/or the textbook.		

LEARNING MAP 3.3

Creating an IC Map With Guided Practice

Outcome

Learners will produce an IC map in a collaborative, guided-practice setting.

Assumption

Learners do not engage in new work from just being told what to do or how to do it. They need modeling and demonstration of new behaviors by an instructional guide or facilitator. Thus, the initial construction of an IC map is done with a facilitator guiding the process, working side by side with the learners.

Suggested Time

120 minutes

Materials

1 copy for each participant of
Professional Learning Communities: What Are They and Why Are They Important?

2 copies for each participant of
Handout 3.3, Blank IC Map (1)

Engaging in Learning

1. Distribute two blank IC map forms (Handout 3.3) and the professional learning community (PLC) paper to participants. Request that participants retrieve from their learning notebooks the New Math Program IC map from our last meeting. Solicit their responses to the question: What two structures did we identify for an IC map in our last session (components, variations)?

2. Further, ask for explanations of what the components represent and, likewise, the variations (see definitions in Learning Map 3.2).

   What do you notice about the language of the components and the variations in the New Math Program IC map? (They are stated in behavioral terms, with action words/verbs clearly describing the components and the variations. There should be no passive voice.).

   Further, guide learners in noting that if they put Teacher (identified in the upper left corner, as the person who is taking the action) in front of each of the components and the variations, this will make a statement about what the Teacher is doing with the change/innovation. The IC map is to do just that—clarify who is doing what! We could make a map of what the principal is doing related to the New Math Program, or a map about what students are doing with it. It is easier to understand the information on the map if only one role group's actions are put on one map. For our work today, put "The Principal" in the name blank, for that is the "who" of our map. And this is an Innovation Configuration Map for "Professional Learning Community"—write that on the title.

3. We have descriptive information of research-based PLCs; please locate that material in the paper that was distributed to you. We will work together on making a map of what the principal does to ensure that the PLC is functioning well. Thus, access one of the Blank IC Map forms (Handout 3.3).

   Read the PLC material, feeling free to mark or write on your copy—our goal is to identify what the principal does to ensure that the components of a PLC are in place. After you have read it, we will work together to identify the principal's role in supporting the PLC components that are described in the paper. Before we do that, take a 5-minute break and get ready for hard thinking together.

4. From reading the PLC paper, what would you expect to see the principal doing when he or she is creating and/or maintaining a PLC? Consider the big actions that are being taken, for this will help us in identifying our components. (The five attributes of a PLC are listed early in the paper, and the pages following describe these attributes more fully; these five items are the attributes that the principal is trying to create in the PLC.)

   After participants have identified the five attributes, they will make a statement about the principal's actions related to the attributes, which becomes the component statement that should be written just below the shaded part of the map. Notice on the Handout 3.2 map how the component statement is very brief. each component should start with a verb to describe the action. This is a challenging activity for learners, as they typically do not engage in a great deal of creative writing. Be patient with them, reminding them that the principal is the subject of the sentence (noted in the upper left corner), the "doer" of the action, and the verb in the component tells the action. Remind participants to keep the component statement crisp. We do not use verbs such as *know* or *understand,* for the verb must denote observable action. Work with them as a group on this, providing guidance but also asking for their input.

   Examples of these statements might be, but are not restricted to, the following:

The Principal . . .

- Supports, develops, and shares leadership with others.
- Guides staff in identifying the intentional learning they will do with colleagues.
- Encourages the articulation of shared values and vision.
- Develops supportive conditions, including structures and strategies (physical and logistical), needed to meet relational conditions

(respect, positive regard, trust) for achieving productive work and learning.

- Encourages staff to share their practices and to give feedback to each other.

They should write these component statements on their map at the appropriate place. Celebrate with another short break when the identification and articulation of components are accomplished!

5. Now comes the development of the variations for each component. Refer back to the New Math IC map (Handout 3.2), and remind participants that the left cell under each component is Variation 1 and is a description of the ideal manner in which the component is being implemented, with the succeeding variations across the continuum being descriptions of decreasing quality.

6. For Component 1, "Supports, develops, and shares leadership with others," Variation 1 might look like this (remember that each phrase should begin with a verb):

Demonstrates responsibility for leading the work and development of the PLC; promotes the identification and assignment of leadership roles for others; develops suggestions for creating leadership in others; shares authority, power, and decision making with others.

Variation 2, then, might eliminate the least important factor from the list above and would look like this:

Demonstrates responsibility for leading the work and development of the PLC; promotes the identification and assignment of leadership roles for others; shares authority, power, and decision making with others.

Variations 3 and 4 might further eliminate additional factors or modify the factors so that the factor loses quality and power.

Variation 5 might indicate doing none of these factors and might be stated like this:

Fails to support, develop, or share leadership with others.

The variation statements are judgment calls and depend on the IC map creator's intentions about what is included in the #1 description, what might be expccted to be seen in schools where PLCs are being created in their ideal form, with the succeeding variations on the continuum expressing lesser value. Both the components and the variations (which expand on the component and give more description) are the product of the map writers and how they decide to define the new practice, program, or process.

There is no magic number of variations, and the same number of variations is not required for all the components on the map. The number of variations used depends simply on the number of descriptors that the map makers decide to provide, based on their predictions about how implementers will put the new practice into their classrooms.

7. With your table group, find the statement of the second component of the PLC that we identified earlier: "Guides staff in identifying the intentional learning they will do with colleagues."

   Write this component statement on your IC map (you have two copies of the map, for marking, messing, redoing, etc.).

   On the back of your map, with your colleagues, generate variations for this component, beginning with Variation 1, which should be a fully described ideal. Remember that the subject is the principal, and when you place that in front of the variations that begin with a verb, you make a complete sentence. After you have your Variation 1, then the task is to use some way to diminish the quality of the variation across the horizontal continuum. Let us use 20 minutes for this work, then we will check with each other.

   As the groups report their Variation 1, ensure that they use verbs liberally to show action, that their descriptions are crisp yet pithy, and that these are reasonable and doable. Many new map makers are prone to use the passive voice instead of the active voice. It is nearly impossible to write in the passive voice if the writers remember who their subject is and start the variations with a verb.

   If an Elmo projector is available, then the maps under creation can be placed on the stage and projected onto a screen for all to see. Invite each group to show their Variation 1 and its related variations across the continuum, looking for verbs and diminishing quality.

8. Similarly, engage the groups in the third component's variations, remembering this is tedious work, and word choice is the hard part of the task. The groups may need to stand and do a Conga line to some snappy music in order to loosen and relax. Use as much positive reinforcement as possible while maintaining fidelity to the process.

   After checking the work on the third component's variations, congratulate the groups for their hard work and emerging product. Request for the next session that they bring good, clear descriptions of a new practice or program that their district or school is adopting and implementing (assuming that the participants are from the same school or district). The idea for the next session is to provide a setting for their independent work, but with several other participants involved in the same change/innovation. It is imperative that they bring this material to work on a product that they will use in their schools.

An example of an IC map of research-based PLCs may be found in Hord and Tobia (2012, pp. 48–53). This map's subject is a member of the PLC, but it easily communicates about the structures of components and variations.

PROFESSIONAL LEARNING COMMUNITIES: WHAT ARE THEY AND WHY ARE THEY IMPORTANT?

Introduction

In education circles, the term learning community has become commonplace. It is being used to mean any number of things, such as extending classroom practice into the community; bringing community personnel into the school to enhance the curriculum and learning tasks for students; or engaging students, teachers, and administrators simultaneously in learning—to suggest just a few.

This paper focuses on what Astuto and colleagues (1993) label the professional community of learners, in which the teachers in a school and its administrators continuously seek and share learning and then act on what they learn. The goal of their actions is to enhance their effectiveness as professionals so that students benefit. This arrangement has also been termed communities of continuous inquiry and improvement.

As an organizational arrangement, the professional learning community is seen as a powerful staff development approach and a potent strategy for school change and improvement. Thus, persons at all levels of the educational system concerned about school improvement—state department personnel, intermediate service agency staff, district and campus administrators, teacher leaders, key parents and local school community members—should find this paper of interest.

This paper represents an abbreviation of Hord's review of the literature (1997), which explored the concept and operationalization of professional learning communities and their outcomes for staff and students.

The Beginnings of Professional Learning Community

During the eighties, Rosenholtz (1989) brought teachers' workplace factors into the discussion of teaching quality, maintaining that teachers who felt supported in their own ongoing learning and classroom practice were more committed and effective than those who did not receive such confirmation. Support by means of teacher networks, cooperation among colleagues, and expanded professional roles increased teacher efficacy in meeting students' needs. Further, Rosenholtz found that teachers with a high sense of their own efficacy were more likely to adopt new classroom behaviors and also more likely to stay in the profession.

McLaughlin and Talbert (1993) confirmed Rosenholtz's findings, suggesting that when teachers had opportunities for collaborative inquiry and the learning related to it, they were able to develop and share a body of wisdom gleaned from their experience. Adding to the discussion, Darling-Hammond (1996) cited shared decision making as a factor in curriculum reform and the transformation of teaching roles in some schools. In such schools, structured time is provided for teachers to work together in planning instruction, observing each other's classrooms, and sharing feedback. These and other attributes characterize professional learning communities.

Attributes of Professional Learning Communities

The literature on professional learning communities repeatedly gives attention to five attributes of such organizational arrangements:

1. supportive and shared leadership,
2. intentional collective learning,
3. shared values and vision,
4. supportive conditions, and
5. shared personal practice.

Each of these is discussed briefly in this paper.

Supportive and Shared Leadership

The school change and educational leadership literatures clearly recognize the role and influence of the campus administrator (principal, and sometimes assistant principal) on whether change will occur in the school. It seems clear that transforming a school organization into a learning community can be done only with the sanction of the leaders and the active nurturing of the entire staff's development as a community. Thus, a look at the principal of a school whose staff is a professional learning community seems a good starting point for describing what these learning communities look like and how the principal "accepts a collegial relationship with teachers" (D. Rainey, personal communication, March 13, 1997) to share leadership, power, and decision making.

Lucianne Carmichael, the first resident principal of the Harvard University Principal Center and a principal who nurtured a professional community of learners in her own school, discusses the position of authority and power typically held by principals, in which the staff views them as all-wise and all-competent (1982). Principals have internalized this "omnicompetence," Carmichael asserts. Others in the school reinforce it, making it difficult for principals to admit that they themselves can benefit from professional development opportunities, or to recognize the dynamic potential of staff contributions to decision making. Furthermore, when the principal's position is so thoroughly dominant, it is difficult for staff to propose divergent views or ideas about the school's effectiveness.

Carmichael proposes that the notion of principals' omnicompetence be "ditched" in favor of their participation in their own professional development. Kleine-Kracht (1993) concurs and suggests that administrators, along with teachers, must be learners too, "questioning, investigating, and seeking solutions" (p. 393) for school improvement. The traditional pattern that "teachers teach, students learn, and administrators manage is completely altered. . . . [There is] no longer a hierarchy of who knows more than someone else, but rather the need for everyone to contribute" (p. 393).

This new relationship forged between administrators and teachers leads to shared and collegial leadership in the school, where all grow professionally and learn to view themselves (to use an athletic metaphor) as "all playing on the same team and working toward the same goal: a better school" (Hoerr, 1996, p. 381).

Louis and Kruse (1995) identify the supportive leadership of principals as one of the necessary human resources for restructuring staff into school-based professional communities. The authors refer to these principals as "post-heroic leaders who do not view themselves as the architects of school effectiveness" (p. 234). Prestine (1993) also defines characteristics of principals in schools that undertake school restructuring: a willingness to share authority, the capacity to facilitate the work of staff, and the ability to participate without dominating.

Sergiovanni explains that "the sources of authority for leadership are embedded in shared ideas" (1994b, p. 214), not in the power of position. Snyder, Acker-Hocevar, and Snyder (1996) assert that it is also important that the principal believe that teachers have the capacity to respond to the needs of students, that this belief "provides moral strength for principals to meet difficult political and educational challenges along the way" (p. 19). Senge (quoted by O'Neil, 1995) adds that the principal's job is to create an environment in which the staff can learn continuously; "[t]hen in turn, . . . the job of the superintendent is to find principals and support [such] principals" (p. 21) who create this environment.

An additional dimension, then, is a chief executive of the school district who supports and encourages continuous learning of its professionals. This observation suggests that no longer can leaders be thought of as top-down agents of change or seen as the visionaries of the corporation; instead leaders must be regarded as democratic teachers.

Intentional Collective Learning

In 1990, Peter Senge's book *The Fifth Discipline* arrived in bookstores and began popping up in the boardrooms of corporate America. Over the next year or so, the book and its description of learning organizations, which might serve to increase organizational capacity and creativity, moved into the educational environment. The idea of a learning organization "where people continually expand their capacity to create the results they truly desire, where new and expansive patterns of thinking are nurtured, where collective aspiration is set free, and where people are continually learning how to learn together" (p. 3) caught the attention of educators who were struggling to plan and implement reform in the nation's schools. As Senge's paradigm shift was explored by educators and shared in educational journals, the label became learning communities.

In schools, the learning community is demonstrated by people from multiple constituencies, at all levels, collaboratively and continually learning and working together (Louis & Kruse, 1995). Such collaborative work is grounded in what Newmann (reported by Brandt, 1995) and Louis and Kruse label reflective dialogue, in which staff conduct conversations about students and teaching and learning, identifying related issues and problems. Griffin (cited by Sergiovanni, 1994a, p. 154) refers to these activities as inquiry, and believes that as principals and teachers inquire together they create community. Inquiry helps them to overcome chasms caused by various specializations of grade level and subject matter. Inquiry forces debate among teachers about what is important. Inquiry promotes understanding and appreciation for the work of others. . . . And inquiry helps principals and teachers create the ties that bond them together as a special group and that bind them to a shared set of ideas. Inquiry, in other words, helps principals and teachers become a community of learners.

Participants in such conversations learn to apply new ideas and information to problem solving and therefore are able to create new conditions for students. Key tools in this process are shared values and vision; supportive physical, temporal, and social conditions; and a shared personal practice. We will look at each of these in turn.

Shared Values and Vision

"Vision is a trite term these days, and at various times it refers to mission, purpose, goals, objectives, or a sheet of paper posted near the principal's office" (Isaacson & Bamburg, 1992, p. 42). Sharing vision is not just agreeing with a good idea; it is a particular mental image of what is important to an individual and to an organization. Staff are encouraged not only to be involved in the process of developing a shared vision but also to use that vision as a guidepost in making decisions about teaching and learning in the school (ibid.).

A core characteristic of the vision is an undeviating focus on student learning, maintains Louis and Kruse (1995), in which each student's potential achievement is carefully considered. These shared values and vision lead to binding norms of behavior that the staff supports.

In such a community, the individual staff member is responsible for his/her actions, but the common good is placed on a par with personal ambition. The relationships between individuals are described as caring. Such caring is supported by open communication, made possible by trust (Fawcett, 1996).

Supportive Conditions

Several kinds of factors determine when, where, and how the staff can regularly come together as a unit to do the learning, decision making, problem solving, and creative work that characterize a professional learning community. In order for learning communities to function productively, the physical or structural conditions and the human qualities and capacities of the people involved must be optimal (Boyd, 1992; Louis & Kruse, 1995).

Physical conditions. Louis and Kruse identify the following physical factors that support learning communities: time to meet and talk, small school size and physical proximity of the staff to one another, interdependent teaching roles, well-developed communication structures, school autonomy, and teacher empowerment. An additional factor is the staff's input in selecting teachers and administrators for the school, and even encouraging staff who are not in tune with the program to find work elsewhere.

Boyd presents a similar list of physical factors that result in an environment conducive to school change and improvement: the availability of resources; schedules and structures that reduce isolation; policies that encourage greater autonomy, foster collaboration, enhance effective communication, and provide for staff development. Time is clearly a resource: "Time, or more properly lack of it, is one of the most difficult problems faced by schools and districts" (Watts & Castle, 1993, p. 306). Time is a significant issue for faculties who wish to work together collegially, and it has been cited as both a barrier (when it is not available) and a supportive factor (when it is available) by staffs engaging in school improvement.

People capacities. One of the first characteristics cited by Louis and Kruse (1995) of individuals in a productive learning community is a willingness to accept feedback and to work toward improvement. In addition, the following qualities are needed: respect and trust among colleagues at the school and district level, possession of an appropriate cognitive and skill base that enables effective teaching and learning, supportive leadership from administrators and others in key roles, and relatively intensive socialization processes.

Note the strong parallel with the people or human factors identified by Boyd (1992): positive teacher attitudes toward schooling, students, and change; students' heightened interest and engagement with learning (which could be construed as both an outcome and an input, it seems); norms of continuous critical inquiry and continuous improvement; a widely shared vision or sense of purpose; a norm of involvement in decision making; collegial relationships among teachers; positive, caring student-teacher-administrator relationships; a sense of community in the school; and two factors beyond the school staff—supportive community attitudes and parents and community members as partners and allies.

Boyd (1992) points out that the physical and people factors are highly interactive, many of them influencing the others. Boyd and Hord (1994) clustered the factors into four overarching functions that help build a context conducive to change and improvement: reducing staff isolation, increasing staff capacity, providing a caring and productive environment, and improving the quality of the school's programs for students.

Shared Personal Practice

Review of a teacher's behavior by colleagues is the norm in the professional learning community (Louis & Kruse, 1995). This practice is not evaluative but is part of the "peers helping peers" process. Such review is conducted regularly by teachers, who visit each other's classrooms to observe, script notes, and discuss their observations with the visited peer. The process is based on the desire for individual and community improvement and is enabled by the mutual respect and trustworthiness of staff members.

Wignall (1992) describes a high school in which teachers share their practice and enjoy a high level of collaboration in their daily work life. Mutual respect and understanding are the fundamental requirements for this kind of workplace culture. Teachers find help, support, and trust as a result of developing warm relationships with each other. "Teachers tolerate (even encourage) debate, discussion and disagreement. They are comfortable sharing both their successes and their failures. They praise and recognize one another's triumphs, and offer empathy and support for each other's troubles" (p. 18). One of the conditions that supports such a culture is the involvement of the teachers in interviewing, selecting, and hiring new teachers. They feel a commitment to their selections and to ensuring the effectiveness of the entire staff.

One goal of reform is to provide appropriate learning environments for students. Teachers, too, need "an environment that values and supports hard work, the acceptance of challenging tasks, risk taking, and the promotion of growth" (Midgley & Wood, 1993, p. 252). Sharing their personal practice contributes to creating such a setting.

Summary of Attributes

Reports in the literature are quite clear about what successful professional learning communities look like and act like. The requirements necessary for such organizational arrangements include the following:

- The collegial and facilitative participation of the principal, who shares leadership—and thus, power and authority—through inviting staff input in decision making
- A shared vision that is developed from staff's unswerving commitment to students' learning and that is consistently articulated and referenced for the staff's work
- Collective learning among staff and application of that learning to solutions that address students' needs
- The visitation and review of each teacher's classroom behavior by peers as a feedback and assistance activity to support individual and community improvement
- Physical conditions and human capacities that support such an operation

Outcomes of Professional Learning Communities for Staff and Students

What difference does it make if staff are communally organized? What results, if any, might be gained from this kind of arrangement? An abbreviated report of staff and student outcomes in schools where staff are engaged together in professional learning communities follows. This report comes from the summary of results included in the literature review noted above (Hord, 1997, p. 27).

For staff, the following results have been observed:

- Reduction of isolation of teachers
- Increased commitment to the mission and goals of the school and increased vigor in working to strengthen the mission
- Shared responsibility for the total development of students and collective responsibility for students' success
- Powerful learning that defines good teaching and classroom practice and that creates new knowledge and beliefs about teaching and learners
- Increased meaning and understanding of the content that teachers teach and the roles they play in helping all students achieve expectations
- Higher likelihood that teachers will be well informed, professionally renewed, and inspired to inspire students
- More satisfaction, higher morale, and lower rates of absenteeism
- Significant advances in adapting teaching to the students, accomplished more quickly than in traditional schools
- Commitment to making significant and lasting changes
- Higher likelihood of undertaking fundamental systemic change (p. 27)

For students, the results include the following:

- Decreased dropout rate and fewer classes "skipped"
- Lower rates of absenteeism
- Increased learning that is distributed more equitably in the smaller high schools
- Greater academic gains in math, science, history, and reading than in traditional schools
- Smaller achievement gaps between students from different backgrounds (p. 28)

For more information about these important professional learning community outcomes, please refer to the literature review (Hord, 1997).

In Conclusion

If strong results such as the above are linked to teachers and administrators working in professional learning communities, how might the frequency of such communities in schools be increased? A paradigm shift is needed both by the public and by teachers themselves, about what the role of teacher entails. Many in the public and in the profession believe that the only legitimate use of teachers' time is standing in front of the class, working directly with students. In studies comparing how teachers around the globe spend their time, it is clear that in countries such as Japan, teachers teach fewer classes and use a greater portion of their time to plan, confer with colleagues, work with students individually, visit other classrooms, and engage in other professional development activities (Darling-Hammond, 1994, 1996). Bringing about changes in perspective that will enable the public and the profession to understand and value teachers' professional development will require focused and concerted effort. As Lucianne Carmichael has said, "Teachers are the first learners." Through their participation in a professional learning community, teachers become more effective, and student outcomes increase—a goal upon which we can all agree.

Credits and Disclaimers

Issues . . . About Change is published and produced quarterly by Southwest Educational Development Laboratory (SEDL). This publication is based on work sponsored by the Office of Educational Research and Improvement, U.S. Department of Education under grant number RJ96006801. The content herein does not necessarily reflect the views of the department or any other agency of the U.S. government or any other source. Available in alternative formats.

SEDL is located at 4700 Mueller Blvd., Austin, Texas 78723; (512) 476-6861. SEDL is an Equal Employment Opportunity/Affirmative Action Employer and is committed to affording equal employment opportunities to all individuals in all employment matters.

HANDOUT 3.3

Blank IC Map (1)

Principal ______________________________

Innovation Configuration Map for ______________________________

Component 1:				
1	*2*	*3*	*4*	*5*
Component 2:				
1	*2*	*3*	*4*	*5*
Component 3:				
1	*2*	*3*	*4*	*5*
Component 4:				
1	*2*	*3*	*4*	*5*
Component 5:				
1	*2*	*3*	*4*	*5*

LEARNING MAP 3.4

Developing an IC Map Independently

Outcome

Learners will produce an IC map of their change, using the skills developed from the previous sessions and working with a small collaborative group.

Assumption

After (1) hearing or reading how, followed by (2) modeling and/or demonstrations, learners (3) engage in guided practice to develop new knowledge and skills in creating an IC map. As a result of the guided practice, learners are prepared to (4) engage in independent practice in making a map—in this case, with a small cooperating group. This sequence of four steps or stages takes novice learners to the emerging expert status.

Suggested Time

120 minutes

Materials

Participants' descriptions of their innovation/change that they are preparing to implement

All the materials of the Chapter 3 Learning Maps, or reference by the IC map makers

2 copies for each participant of
Handout 3.4, Blank IC Map (2)

Engaging in Learning

1. Organize participants into three groups, each of which will focus on one of the following questions:
   - Why do we require precision in articulating the change/innovation that we are adopting?
   - What are the two major constructs of an IC map and their definitions?
   - How do we create an IC map (briefly), enumerating the steps of the process?

   Explain that a crisp but pithy review of the process of constructing IC maps will be helpful for our work today.

   To that end, each of the groups will prepare to instruct the large group about one of the three items noted above. Assign one of the topics to each small group, reminding them that in their materials for the past several sessions, we have information and activities that aided us in reaching the goal or outcome of the specific lesson.

Each group is to prepare a lesson for the large group that answers the question they have been assigned. The first question group might refer back to Learning Map 3.1 to support their lesson plan; this group will have 5 minutes for their teaching. The second question group could refer to Learning Map 3.2 for planning assistance; this group will have 10 minutes. And the third group could refer to Learning Map 3.3; this group will have 15 minutes. Are there questions about your tasks?

You have 15 minutes to prepare your lesson for the large group.

2. Let us now conduct our lessons, in the sequence noted above. Participants should feel free to ask the "teachers" questions, or express needs for clarification. After each session, ask the large group about questions and/or needs for clarification they might have. Respond as appropriate.

3. Take a 10-minute break, then ask participants to begin their task of constructing an IC map of the innovation on which they have all determined to work. They are to do this with their small group, so interaction should be noted during this process. They should use the descriptions of the innovation that they have determined to focus on and that they have brought. Inform the groups that after 20 minutes, they will report on their progress of articulating the components of their change/innovation. Their first task is to determine the person or role of person for whom the map is being constructed (e.g., teacher, principal, other), and fill in the Name/Role section on Handout 3.4 accordingly. Tour the tables in order to be available to give help and support or to ask questions that will assist the small groups to work more meaningfully.

4. Support each group in presenting their components to the large group. Encourage other participants to provide warm feedback (positive, acknowledging a favorable aspect of the list of components) to the presenters as well as cool feedback (questions or comments that stimulate the map creators to rethink their components or the language of the statements of the components, for example: Is each component stated as an action, with a verb? Is each component stated so that the reader gains a mental picture of what is happening vis-à-vis the statement?). Again, creating the components is challenging work—to identify the major pieces of the innovation and to state them in brief, action-oriented statements. Thus, congratulate and celebrate their work.

5. Now comes the development of the variations for the components. Direct participants to go back to Learning Map 3.3, Steps 5–7, and use these steps as directions for developing their variations. Solicit questions. Ask the groups to create variations for the first component on their map.

   Ask groups to turn to another group and pair-share—that is, review one team's work, giving warm and cool feedback, followed by the other team doing the same. After these conversations, ask a group to share their work; request comments or questions from the large group.

6. Groups should complete their maps during the remainder of the session. If they have not finished, ask for their completion 4 days hence, and ask each group to email their map to the facilitator for review before the next session.

   In any case, distribute chocolate to reward participants for their hard work, or invite everyone to the local beverage depot to review the day's triumphs and tribulations.

HANDOUT 3.4

Blank IC Map (2)

Name/Role____

Innovation Configuration Map for ________________________________

Component 1:				
1	*2*	*3*	*4*	*5*
Component 2:				
1	*2*	*3*	*4*	*5*
Component 3:				
1	*2*	*3*	*4*	*5*
Component 4:				
1	*2*	*3*	*4*	*5*
Component 5:				
1	*2*	*3*	*4*	*5*

LEARNING MAP 3.5

Reviewing and Revising the Map

Outcome

Learners will produce a reviewed and revised edition of their developing IC map.

Assumption

A considerable amount of time, effort, thoughtfulness, trial, and error are required for constructing a useful IC map. This task is typically not done well by a single individual; a small group that includes at least one person with clear knowledge about the innovation is necessary. Further, careful reviewing and revising during map creation will increase the accuracy of the map.

Suggested Time

60 minutes

Materials

1 copy of each small group's IC map for all of the groups

1 copy for each participant of
Handout 3.5, Critique of an IC Map

Engaging in Learning

1. Organize so that two groups (of map makers) meet together. Each group will independently review the other's map and mark items that are not clear, are questionable, or need clarification. Distribute Handout 3.5, Critique of an IC Map.

   The reviewers should look for active voice, marking passive voice phrases; they should also check to ascertain that the person (identified in the upper left corner of the handout) is understood to be the subject of the component and of the variation phrases that begin with a verb (thus providing a complete thought and sentence).

2. After the critiquing and marking are completed, each of the paired groups should plan for how they will share their feedback with the other group (note hints about feedback, provided on Handout 3.5). Then the two groups share with each other.

3. Solicit questions and/or concerns across the large group. Ask for volunteers to share their maps and the feedback they received.

4. Request that the groups give further critical attention to their maps, revise their maps, and produce clean copies of their maps to bring to the next session.

HANDOUT 3.5

Critique of an IC Map

Read the entire map, thoughtfully reading each component and its variations.

Mark any words or phrases that are not clear or that do not make sense to the reader. This includes marking anything that is questionable or needs clarification.

Review the document for active voice—that is, put the person (identified in the upper left corner) of the map in front of each component and in front of each variation to check that these produce sentences and complete thoughts.

Review the variations on each continuum to assess their diminishing quality from left to right.

Review your markings and consider how you will provide feedback to the authors of the map or how you will construct questions for inquiring about items that are not clear.

Remember that feedback is most comfortably received when a warm (glow) item is offered (and congratulated, possibly) and then cool (grow) comments/ questions are provided.

The purpose of this activity is the improvement of the map.

LEARNING MAP 3.6

Field-Testing and Revising the Map

Outcome

Learners will produce a field-tested and revised map.

Assumption

It is imperative that the IC map is clearly drawn so that another person (not the map maker) can devise a mental image of the innovation in action, in its intended setting. Clarity is important so that the reader of this authentic map gains understanding of *who* is doing *what*, vis-à-vis the map.

Suggested Time

60 minutes

Materials

Copies for each group of all the revised maps for all groups, developed from the work of the previous session

1 copy for each participant of
Handout 3.6, Conducting the IC Conversation

Engaging in Learning

1. Check to ensure that each group has copies of its revised IC map. Solicit questions or concerns from the group about their map or the map-making process.
2. Ask each group to identify two to three individuals who are using the innovation or expect to use the innovation.
3. Share Handout 3.6 with the participants, asking them to review the text. Review each item with them, soliciting questions, and asking the participants questions to ensure they understand the directions for the conversations they will have with the users of their innovation that they identified.
4. Direct the groups to practice the interviews with their group members, with one member interviewing, the second member serving as the interviewee, and the third as an observer to give feedback. If, in the interview, the interviewee has not mentioned all the components on the team's IC map, then inquire about them. After this role-play, ask for comments across the large group and have participants share their reactions.

5. Change roles within the teams so that each member practices the interviewer role. At the conclusion of these practices, solicit questions, reactions, comments.

6. Request that the small groups make plans for their interviews.

   They might determine to have the same person on their team interview all the interviewees so as to keep the interviewer factor constant, or they may opt to assign each team member to an interviewee.

   After they have completed all interviews, they should convene their team to analyze and synthesize their learning from the interviews, giving attention to any needs for adjusting wording or other factors, and then revising the IC map.

7. Direct the teams to polish their maps and bring revised copies for each of the teams to the next session.

HANDOUT 3.6

Conducting the IC Conversation

- Schedule a 30-minute visit with each of the interviewees, making certain that the date and time are agreeable with the interviewee.
- Identify a location for the conversation, considering any noise and visual distractions.
- Greet the interviewee outside the interview location, exchanging comments (about the weather, last night's football game, etc.) as the interviewer leads the interviewee into the space for the conversation. This is to initiate a friendly atmosphere, as some individuals may be anxious about this activity.
- Explain the purpose: to try out a process for helping individuals understand *what* any new practice, program, or new process looks like when it has been implemented in its desired setting. Emphasize that you are deeply grateful for their time and involvement, since it will help you know if the process is on target, and that changes will be made dependent on the conversations that you have.
- Explain that, because we will find their comments valuable and we don't want to get different persons' comments jumbled, you will take notes on your computer (if this is comfortable for you to do; or you could use pen and paper or a tape recorder, if that is available and the interviewee is agreeable—ask for permission to tape record if that is to be your method of keeping notes).
- Initiate the conversation about the map topic by using open-ended questions, such as the following:

  Tell me, if you will, please, about the new . . . (whatever your IC map focuses on)? What would I see you doing? If I were the proverbial fly on the wall, what would I observe?

- Receive any comments without judging; if more description or explanation is needed, ask the interviewee to

  Help me understand . . . or
  Could you give me an example of . . .

  (You are looking for enough detail to be able to mark a variation that describes this user of the program.)

- If the interviewee has not commented on specific components, ask about those:

  Could you tell me about how you work with the assessments of the program? (or whatever the missing components might be—check for each of the unmentioned components individually)

- Summarize briefly what was communicated and thank the person abundantly, emphasizing his or her contribution to the effort.

LEARNING MAP 3.7

Sharing the Map With Implementers

Outcome

Learners will create a plan for sharing the IC map with implementers.

Assumption

Like any new item's expected use, just giving the IC map to those expected to use it will not be very effective. Learning about the map, its parts, its active orientation, and its components and their variations requires a facilitator who will carefully craft a plan for teaching the implementers *what* the map is, *why* we have it, and *how* to use it.

Like its predecessors, this lesson for leaders is vital to ensure that the map is well used and results in impact on educators and, subsequently, on students.

Suggested Time

60 minutes

Materials

Copies of each group's field-tested and revised maps (resulting from the work of the previous session) for all of the groups

1 copy for each participant of

Handout 3.2, IC Map for the New Math Program
Handout 3.7, Introducing and Using the IC Map

Engaging in Learning

1. Request that learners, in their groups, access the IC maps of each of the groups. Read/review each of the maps with half-closed eyes—as you read each map, can you envision the action that is occurring? The map should give us a mental picture of action related to the innovation. Solicit volunteers to describe pictures that they "see" from one of the maps, sharing the text that stimulates the mental picture.

   This is one of the steps you will use when sharing the map with potential implementers in the innovation's expected setting. Note that it is unlikely that anyone will be demonstrating the #1 or ideal variation at its introduction (although this is possible, of course), but the goal is to support implementers until they do reach ideal practice, vis-à-vis Variation 1.

2. For introducing the map, we will use Handout 3.7, Introducing and Using the IC Map. Access this handout and give it a careful read. But also have your finalized copy of your map available.

3. Reconvene the group, and review with them the points made on Handout 3.7. Solicit the group's questions, concerns, or needs for clarification about the introduction to the map and the discussion of its various uses (of which there are more, but these four are sufficient at this time). Include the participants in responding to any queries.

   The task now is to create a plan for how you will introduce your map to the relevant group(s). You may use any format, but please make it clear, simple, direct, and uncomplicated. Clearly mark the steps that you will take and how you will take them. You might use the information on Handout 3.7 to guide your communication plan.

   Work as a team to design your plan. Note the amount of time that you believe will be required for each step; this will prevent you from trying to do too much in one session. You could design your plan for two sessions if that seems a good idea and is possible.

   When the plans are completed, mount them on the wall around the room and conduct a Gallery Tour—everyone parades around to review all the plans.

You and your colleagues have addressed a very large challenge: the constructing and sharing of an IC map with implementers who will use it. Adjourn the session to reconvene at Le Gran Pavillione for a *Bravo!* celebration.

HANDOUT 3.7

Introducing and Using the IC Map

Introducing the Map

[Use this text to guide your work in introducing the map to a group.]

Directions that you may use with your map and its implementers follow.

First, notice several structural elements on the IC map:

- In the upper left corner is the name of the person or role of the persons who are the "actors" of this map; that is, they are taking the action.
- Notice that each piece of text on the page begins with a verb to denote the action that the actor is taking; therefore, if you mentally place the actor in front of each of the action phrases of text, a complete sentence results.

Second, the map is constructed of components and the components' variations.

- You will find the component stated at the top of each row of cells. Please see Handout 3.2, IC Map for the New Math Program. Component 1 of that map is as follows:

Component 1: Selects Objectives

(1)	*(2)*	*(3)*	*(4)*
Selects objectives, in sequence from the district list, and may add objectives to address the needs of particular students.	Identifies objectives from other published documents that cover the district list.	Refers to other sources for objectives not related to the district list.	

How many components do you see on the map?

What do you think the components represent?

The teacher is the actor on this map (noted in the upper left corner of Handout 3.2). If you combine "The teacher" with Component 1, a sentence is created: "The teacher selects objectives." This sentence is very brief, crisp, and not descriptive—there are variations to do the describing.

For the first component, there are three variations, or descriptions, of what we might observe in a classroom where the teacher is implementing the New Math Program. The ideal variation, best practice, or desired picture of what the teacher is doing is Variation 1. It is this variation that is the ultimate goal of teacher's actions for implementation of the innovation. Moving along the continuum from #1 are reduced, or diminished-in-value variations—these might be expressed by teachers as they are learning to use the New Math Program.

Please note that each component does not have the same number of variations. Variations are based on predictions of what any actor is likely to be doing with the innovation as he or she is learning to use it.

Now that you have a sense of how the map is constructed, and of what, read the map in its entirety. Close your eyes. Do you get a mental picture or a general sense of what the teacher is expected to be doing? Does this map give you a general sense of what the New Math Program is?

You could use these steps in introducing your map by substituting the New Math Program with your map's name and information, or creating your own approach, using the New Math as a guide.

Using the Map

There are multiple uses for the IC map:

- Describing the innovation—The map can provide a good picture of what the innovation is and how it should be used. And it provides a common language by which everyone involved can dialogue and discuss the new practice.
- Setting goals—As implementation progresses, interim goals can be established by asking implementers to achieve a particular level of variation (e.g., by Thanksgiving break, everyone will have progressed and be operating at Variation 3 on each component). Establishing such a goal gives the implementers the chance to access support and to learn more; then the goal can be moved to operating at Variation 2, and so on. Some change leaders use the components as "goal posts." That is, they may ask implementers to work on Components 1, 2, and 3, and ignore Component 4 until later.
- Checking progress—In addition to describing the innovation and setting goals for achieving its use, probably the most significant use of the map is for determining progress of the user/implementer in order to provide assistance or support. We would unequivocally state that in this case, the map is a growth-inducing (learning) instrument. It is not meant to evaluate implementers as they work to learn about the innovation and put it into practice. But it gives both the implementer and the change leaders, or coaches, specific data so that appropriate support and assistance may be provided. The map is designed to engage each implementer, communicating what he or she must learn do to in order to meet the goals and to use the innovation in a high-quality way.
- Checking progress—The map, as a result of determining progress, can serve as a formative assessment tool by the change facilitators/leaders/coaches to record and report the progress of implementation of individuals, groups, or an entire organization.

This introduction of the IC map and its use may be employed by principals, teacher leaders, and others for introducing any IC map to any group that will be responsible for using the innovation that it describes.

LEARNING MAP 3.8

Using an IC Map for Developing an Implementation Plan

Outcome

Learners will identify how the IC map can be used to initiate planning for implementing the change.

Assumption

A traveler does not usually depart in an automobile, or airplane, or ocean liner without a clear identification of the desired destination and how to reach it. Likewise, it is imperative that change leaders know precisely where or what the target "destination" is, and make appropriate plans to reach the goal.

Suggested Time

60–90 minutes

Materials

1 copy for each participant of
Handout 3.8, Designing an Implementation Plan
Handout 3.2, IC Map for the New Math Program

Engaging in Learning

1. We will focus on the creation of a plan for implementation, considering the six strategies, which we learned about in Learning Map 2.1. Please organize yourselves around tables in groups of three to five individuals. Write the six strategies in your journal or on a blank sheet of paper. Consult with your tablemates to complete your list, and refer to your material from Learning Map 2.1 and Handout 2.1.
2. Provide 10 to 12 minutes for this task, and then discuss their responses for accuracy:
   Developing and communicating a shared vision of the change
   Planning and providing resources
   Investing in professional development
   Checking progress
   Continuing to give assistance
   Creating an atmosphere/context for change
3. Distribute Handout 3.8. It contains the six strategies that we just recalled. Notice that the wording of the strategies and the order of the strategies (on Handout 3.8) have been changed a bit. We will understand why this is as we do our work.

Now, distribute Handout 3.2, to which we gave attention earlier in our work in learning about IC maps. We will use the strategies framework to initiate our thinking about an Implementation Plan for the New Math Program, whose IC map we have in hand.

We will practice Designing an Implementation Plan, making certain that all strategies are considered. Your task with your tablemates is to study the IC map in order to refamiliarize yourselves with this program, especially noting the components and the variations. Then, making notes on Handout 3.8, suggest how the IC map might be helpful in designing a plan for implementing the New Math Program, and jot down your ideas under each specific strategy.

For example, on Handout 3.8, can the IC map for New Math help with the strategy of Developing and Communicating a Shared Vision for the Change? If so, how? Consult with your table group, jot down suggestions, and we will discuss this first strategy in 5 minutes.

4. Creating the vision has already been accomplished and presented in Handout 3.2. In Learning Maps 3.2–3.6, we experienced and explained creation of an IC map, but we don't have to make a map.

   If we did not have one, then constructing a map would be a first order of business so that it could be used as the written vision of the change.

   The IC map *is* the vision of the change, for it describes *what* will be done by *whom* when using the new practice or innovation. Introducing and initially communicating the vision to implementers was the focus of Learning Map 3.7.

   Since we have the New Math IC map, what remains to be done for the Implementation Plan is the design of actions for reminding all implementers (change leaders, facilitators, and other key individuals) about the innovation and its vision of change. A continuous array of posters, flyers, signage in the lounge/work rooms, various relevant editions of the IC map, notes in the faculty bulletins, brief reports in newsletters to parents and community members—all serve to keep everyone aware of the change and its progress. The IC map should be readily available for reference in designing the communication procedures for sharing the vision. The plan for this communication may be placed with this first strategy, for *creating* the vision and *communicating* about it comprise the strategy.

5. Questions? Reminder: We are studying the IC map for New Math to determine how it might (or might not) aid us in beginning to create an Implementation Plan. Please use this IC map and work with your table group on the Implementation Plan for the next two strategies: Creating a Context for Change, and Planning and Providing Resources. We will have 20 or so minutes to work on this, then share with each other.

6. Call the large group together again, solicit their ideas, inviting reactions to their suggestions from others, and discuss all points offered. Include the ideas below if they have not been offered.

   It is a "sticky" question to determine whether the context strategy should be initiated first—that is, before the vision strategy. Certainly, it will be very helpful if the context is already one in which change is seen as positive and productive, where trust is evident among the staff and respectful relationships abound. However, if the context needs attention, then the vision of the change can be shared if it has been created, and that can contribute to developing an open and collaborative context and provide the focus for collaborative conversations.

It is also possible to work on creating the vision and the context simultaneously, if creation of the vision is done openly and with abundant input and participation.

Sharing the available vision early can contribute to an open, transparent context where all are informed of the innovation and understand clearly what it is and why it is important. Such a context enables trust to develop and commitment made to the vision by all. Developing such a context is supported by communicating clearly what the change is and why it is being adopted. Using the IC map to communicate these messages is another productive use of the map.

Actions that have already been taken to create a positive and receptive atmosphere for the change should be identified and entered on the Implementation Plan. Additional actions to be taken may then be added for consideration across the several years of the implementation.

7. Creating an (implementation) plan and identifying its required resources is our next strategy. Can an IC map contribute to designing this strategy of the Implementation Plan? Yes. Creating that plan is what we are now considering, and we look to the map to ascertain what we will need to do and what resources will be needed to reach the ideal variation of each of the components.

   The IC map—that is, the vision—is being used to guide the development of several strategies of the plan for implementation. Use of the IC map ensures that the plan is on target and supports the introduction and installation of the change/vision in classrooms or wherever its intended location may be. This strategy should result in a list of all actions needed to reach implementation of the map's components and all resources (both human and material) that are needed for using the envisioned change.

8. Investing in professional development—that is, adult learning—for all who are involved in implementation or in supporting it, is our next strategy. Following that are Checking Progress, and Continuing to Give Assistance.

   Please take a break and then continue with your group to work on these three strategies, noting how the IC map can help to develop the strategy or not, and how.

   After the break, reconvene the large group to solicit suggestions about using the IC map for investing in professional development/professional learning. It should be clear that using the IC map for planning professional learning for all implementers and their support personnel is an imperative. The map provides specific descriptions of what the teacher, or principal, or other role groups will be doing when involved with the New Math Program. Thus, the map literally dictates what the initial professional learning should be in order to prepare all for the *what* and *how* of using the New Math, as indicated on the map. If there are those who are already skilled in using the new practice or program, then differentiated professional learning should be provided. Ask the group if anyone needs clarity, respond to those needs, and move on to Checking Progress.

9. Many people who do not know about using the IC map for all the purposes identified above do indeed use the map to check the progress of implementation of the new practices. It sounds much more positive, and less threatening, to use the term *checking progress* (rather than *assessing* or *monitoring*) for the IC map should be thought of as a growth-inducing tool, not as evaluation.

   The map's structure of easily identifiable components with their descriptive variations placed on a diminishing-quality continuum presents this tool as easy to use, providing immediate feedback on the user's implementation of the New Math. When the facilitator engages the user in the discussion of the user's behaviors and where they appear on the map, checking progress is immediate and collaborative, especially when the printed map is in front of them and they are cooperatively marking on it where the implementer is.

   In the Implementation Plan, the frequency of using the map for checking progress (who will do it, how, and where) should be noted in the cell related to that strategy, where you are recording your notes. It is easy to see that the IC map is a very useful item for the strategy of checking progress.

10. Now let us examine our understandings of the Continuing to Give Assistance strategy. What have you determined about this strategy and the possibility of the IC map's support of it? It seems easy to see that checking progress and giving assistance go hand in glove, and essentially are the two components of coaches' work. Checking progress is the data gathering action, which feeds the determination of appropriate assistance to the implementer. While the IC map is used for the checking progress action, it is not strong in its capacity for suggesting assistance, although the facilitator and the implementer may review the map to understand where the implementer is and where progress should be driven.

    The source of assistance comes, in a major way, from the facilitator's knowledge base, experience, and skills in giving feedback, both warm and cool. Much can be gained by the facilitator from exploring the reports of change efforts and reviewing the local school/district's resources in order to discover the various types of assistance available, or possible.

11. Bring closure to the session by sharing that in the next sessions, new concepts and tools will be introduced that will have many applications and may be used to adjust or modify the strategies in the Implementation Plan.

    Remind participants to bring their notebook of materials to the next session, for we will build on what we have already done.

HANDOUT 3.8

Designing an Implementation Plan

Step 1: Developing and Communicating a Shared Vision of the Change	
Step 2: Creating a Context for Change	
Step 3: Planning and Providing Resources	
Step 4: Investing in Professional Development	
Step 5: Checking Progress	
Step 6: Continuing to Give Assistance	

4

Stages of Concern

Understanding Individuals

EVERYWHERE CONSIDERS THE COMPELLING CASE FOR CONCERNS

EveryWhere district now has three carefully constructed IC maps that represent the three innovations for math. These maps give staff a very clear description of the changes and a way to identify progress. The maps also provide staff with the mental images and text pictures of the change as it would appear in their day-to-day work and practice. And while it took a considerable amount of time and effort to create the maps, EveryWhere has roadmaps for their identified changes that are action based. While the team was creating the IC maps, Bob Yurself noticed how knowledge and understanding of the innovations were obviously increasing. He observed that trust appeared to be developing as the group openly discussed ideas, and objections, about the IC maps of the new math practices. And confidence in the work seemed to be growing among the staff members, as they began speaking knowledgeably about the innovations on which they were focusing. When they added more teacher leaders from each of the schools to help create the IC maps, Bob was pleased to see how the new members were welcomed into the group and encouraged to share their thoughts and ideas in designing a map that could work for those implementing it.

As a master gardener, Bob could clearly see how this garden of change was taking shape and becoming more and more a reality. The district now had a clear vision of what would be planted where, and how that planting would take place. When Bob recalled the year he started working on his garden, he remembered how so many people had such different reactions to what was being planted. The neighbor who had a love of flowers felt that Bob's garden had too many native plants. And Bob's wife wanted to have more fresh herbs and vegetables in the garden. And some of the children in the area weren't happy because they were now blocked from using Bob's back lot for getting to one another's homes. Bob could clearly see that changing anything brings about a variety of feelings and attitudes.

Then Bob remembered reading with his staff in their book study on change that while an accurate map is essential for traversing new territory (IC map), there are other factors that have to be considered when guiding and supporting individuals through the process of implementing change. It isn't enough to just have the vision; one also has to consider that each person will process the journey of change with his or her own feelings, reactions, or attitudes (affect) and a continuum of behaviors. It was a reminder to Bob that one size does not fit all. That is why the Concerns-Based Adoption Model (CBAM) tools—Stages of Concern (SoC) and Levels of Use (LoU)—would be so helpful in differentiating the support for the math changes so individual needs and concerns could be met. In this effort, then, everyone in EveryWhere District would receive the kind of support that would be the most helpful to his or her concerns related to the change.

It would be useful, it seemed to Bob, to apply the insights derived from Stages of Concern to the design of the large-group learning sessions about the new math practices. Bob also realized that considering the SoC data about each person would increase the effectiveness of coaching sessions with individuals. In this way, not only would the strategy for professional development benefit from SoC, but so would the checking progress strategy used by coaches (see Chapter 1).

And in using concerns data, yet another strategy would be addressed—that of creating the context for change. When individuals feel that their reaction and feelings are seen and heard, and that the individual is valued, their enthusiasm and energy committed to implementation escalates. Interesting data, reported by Kennedy (2012), from *The MetLife Survey of the American Teacher* (2011), reveal the largest decrease in job satisfaction of teachers in 20 years. However, significantly, Kennedy reports Daniel Pink's (2010) three key elements of motivation that address internal drive and increased job satisfaction:

Autonomy—Individuals want control over their work.

Mastery—Individuals want to improve at what they do.

Purpose—Individuals want to be part of something bigger than they are.

Thus, inviting implementers into the conversation about Innovation Configuration (IC) and SoC data, in order that they elicit suggestions for assisting themselves and colleagues, enables an increase in both autonomy and mastery. In addition, it provides the opportunity to be involved in the larger effort—that of considering and creating authentic support and assistance—and expands the stage upon which the individual implementers are performing.

Like IC, SoC is a valuable tool for use with various strategies.

Bob and his team were ready to move into the next stage of learning that would help them understand how individuals predictably have various Stages of Concern that arise during the implementation of a change and, subsequently, how they should consider this information for inviting implementers into a conversation about their team's SoC data so that they generate ideas for the group's support, thus increasing the pool of collegial facilitators. When the implementers get to play this role, they begin to feel rewarded for their ideas, skills, and professionalism (Hord & Tobia, 2012).

LEARNING MAP 4.1

Considering the Compelling Case for Concerns

Outcome

Learners will explain the concept of Stages of Concern and use individuals' comments to identify their concerns.

Assumption

While an accurate map (IC map) is essential when traversing new territory, there are other factors to be considered when guiding and supporting individuals through the process of implementing change. One of these is the individual's feelings, reactions, or attitudes (affect) about the change—we understand this through the concept of Stages of Concern.

Suggested Time

90–120 minutes

Materials

1 copy for each participant of
Handout 4.1.a, Novice and Experienced Teachers
Handout 4.1.b, Typical Expressions of Concerns
Handout 4.1.c, Practice Scoring Stages of Concern

Engaging in Learning

1. Distribute copies of Handout 4.1.a, Novice and Experienced Teachers, and tell learners that we will hear a report of an early study of teachers' reactions or attitudes. Direct their attention to the handout and the two groups of respondents in a small research study: experienced teachers and novice teachers, noted on the horizontal axis. Direct their attention to the vertical axis on which two questions appear. Ask learners to take notes in the appropriate boxes about the study's findings that they will learn about in the short report.

2. Share the report:

   Quite a number of years ago, an educational researcher conducted a study of two sets of teachers: one set was novice, or new, teachers, just hired for their first year of teaching; the other set was experienced teachers who had been in the classroom for 5 years or more. The two sets of teachers were asked two questions—please see these questions on the handout.

   When the teachers were asked about their concerns about teaching, what did the novices report?

   Yes, they unanimously said that they were mainly concerned about keeping discipline and a quiet classroom. When the experienced teachers were asked this question, their response was, whether their students were learning what was being taught. Did you put those responses in the two cells on your handout?

When the experienced teachers were asked about their satisfaction with teaching, what do you suppose they replied? Yes, they were very pleased and satisfied when their students learned well. When the novice teachers responded to this same question, they stated that the thing that satisfied them was . . . the upcoming holidays! Ah, this makes sense, doesn't it, for new teachers?

The question for us is this: What do we do about planning professional development for these teachers who have widely divergent reactions to the questions? (Solicit responses, driving to the conclusion that we must plan differentiated professional learning for the two sets of teachers, for they have very different concerns about teaching.)

Take a short break, and then we will hear a report of an evaluation study.

3. Dr. Frances Fuller was a professor of educational psychology at The University of Texas/Austin, where she taught the first professional sequence course for potential teachers, and taught it to students in their freshman year. At the end of the course, Dr. Fuller always administered an evaluation of the course. At the end of this particular course, she asked, among other questions, "How relevant is this course to you and your teaching goals?"

   She was stunned at the reactions from that question: 90% of the class reported that the course was totally irrelevant to them. Frances, being the person that she was, chatted with the 10% who found value, and she discovered that these several students (different from the 90%) had very recent teaching activities with students: Sunday School, summer playground crafts and skills development of kids, babysitting, vacation Bible school. Frances hypothesized that their attitudes and reactions to the course were impacted by their knowledge and current, or recent, experiences with students. These reports challenged Frances to learn more about teachers' and student teachers' concerns, which led over time to the concept and measures of Stages of Concern.

4. Access Handout 4.1.b, Typical Expressions of Concerns, so that we may study Frances Fuller's (1969) three-stage articulation of teachers' concerns and Hall, Wallace, and Dossett's (1973) seven stages.

   Notice first that there are seven Stages of Concern, including a zero stage, and that the numerals used to accompany their names are Arabic numerals, to differentiate from the Roman numerals associated with Levels of Use, which we will study in future sessions. Notice also that the numerals begin at the bottom of the column and ascend upward.

   In the far left vertical axis you will see the three stages that were identified by Frances Fuller: Self, Task, Impact. Moving across horizontally for each of these three, please note the expansion of Self into Stages 1 and 2; Task is represented by Stage 3; and Impact is expanded into Stages 4, 5, and 6. Stage 0: Unconcerned, is noted as an Unrelated stage. Hall and colleagues (1973) expanded Fuller's definitions to these seven.

When an individual is at Stage 0: Unconcerned, you would hear him or her say such things as "I don't know anything about the new math" or "I am not interested in learning about the new math, for I have no expectation of being involved with it." The key here is no thinking about, being interested in, or being concerned about the innovation. However, if or when the individual expresses interest in learning about new math, he or she is considered to be at Stage 1: Informational. At Stage 2, the person knows about new math, but is self-focused: Can I do it? What will it mean to my professional and personal life? Will I be comfortable with it?

In all cases, SoC 1 and SoC 2 are Self concerns, as Fuller indicated.

Fuller's Task concerns constitute a large percentage of concerns about a new practice, program, or process, as does the single SoC 3: Management. Management concerns are commonly found when individuals are approaching using the new math or are engaged in using it. Expressions of these individuals very commonly focus on their reactions about time (the shortage of it, or occasionally, too much), materials, lack of being organized, not knowing what to expect next.

There are three Impact concerns: Consequence, Collaboration, and Refocusing—all of which focus not on Self, not on the Tasks, but on increasing the effectiveness of the new math or whatever the innovation may be. SoC 4: Consequence expresses this interest by such comments or feelings as "wondering how the students are accepting the new practice" or, as the chart suggests, "How is my use of the new math affecting kids?" This kind of concern occurs typically after the user has the new way under control and management issues have been solved. It is a time when the implementer is questioning himself or herself and thinking about how this new thing is working for clients.

SoC 5: Collaboration moves beyond the single implementer, as this individual is thinking about how he or she might work with others who are implementing the new math. In this concern, *purpose* is a telling factor, for there may be two very different reasons for which implementers may wish to work together. We frequently see two new math users coming together to plan and to organize, determining that one will prepare lists of resources and acquire them for their lesson plans while the other designs and organizes ways to set up the equipment and materials. The reason for their work together is to conserve time and energy—rather than this being thought of as a Collaboration concern (because the purpose is not to impact their students' learning), it really represents a Management activity, and would thus be thought of as SoC 3, and any facilitating support would be addressed to Management concerns.

The second reason that we see implementers thinking about coming together to collegially explore possibilities, design activities, and recreate structures and strategies is to expand the results of their work for the purpose of increasing students' benefit (that is what Impact means: influencing clients' outcomes). This is the authentic Collaboration that is SoC 5.

And then there is SoC 6: Refocusing, another impact concern, whose motivation is increased results or expanded outcomes for clients. SoC 6 denotes the consideration of a bigger or major adjustment to the new program or practice, or perhaps substitution with an alternative program, which would be seen as superior to new math. The SoC 6 concerns are about how an individual is thinking and expresses ideas "about something that would work even better." It is not often that implementers reach this Stage of Concern, as concerns for self and management keep most implementers working to reduce these concerns.

5. Let's do some practicing in identifying the SoC of individuals by coding their comments. Please find Handout 4.1.c, Practice Scoring Stages of Concern. First, please pair with a colleague to work on this activity together. You will read each statement, underlining any words or phrases that appear to give you clues or hints about the SoC being expressed. You may keep your chart of concerns expressions (Handout 4.1.b) handy for reference. Above each underlining, write the numeral of the Stage of Concern being expressed. Then, in the right column adjacent to the statement, place the numeral and name of the concern you have identified. Let us do the first example together in guided practice.

   Read Statement 1 to the group, asking them to follow on their handout. Tell them that they should underline time, materials, and organizing all things necessary. Call their attention to the Stages of Concern chart (Handout 4.1.b) and Stage 3, noting how the statement matches well with the Statement 1. Thus, place a 3 over the words *time, materials,* and *organizing all things necessary.* Place a 3 in the right column and the name of the stage, Management. Ask for questions, or needs for understanding, or explanations of why we marked this statement as we did.

   Now, work with your partner on Statement 2. Read, mark words, and write the numerals and the name, and then we will check, so hold for our discussion after you have completed this statement.

   Yes, you should have underlined I am studying and asking questions. Now, hold up fingers to show me what stage this indicates. Yes, it is SoC 1: Informational. Questions? Clarifications? Finish the statements as we have done, working with your partner and talking together so that you reach consensus and mark your statements the same. This means I should hear discussion of the markings and commentary about the evidence for marking that way. (Tour the tables while they complete their work.)

   Now that you have completed the statements, let us review your work. I will provide the appropriate responses to you, so that you begin to align your judgments with the "company line."

For Statement 3, you should have marked students need and help students, and possibly more productive experience. Notice that this statement is not about the teacher, but about the students and a factor that is influencing the students. And the teacher is thinking how he or she might fix it a bit to help the students. This statement is coded as 4: Consequence, for the reasons stated.

Questions?

In Statement 4, you should have underlined don't even know. Notice that the individual doesn't express any interest in learning about new math. Score this one a 0: Unconcerned, since there is no indication of knowledge or interest in getting information.

For Statement 5, you should have underlined we need to coordinate and all our kids are able. The person is suggesting that the team leaders get things coordinated for their students' benefit, indicating 5: Collaboration, because of those two factors.

Statement 6 indicates a person with SoC 2: Personal concerns. You should have noted and underlined don't know about, whether I can handle, so different from what I am accustomed to. This person is expressing doubt and lack of personal confidence in using the new history curriculum.

The SoC 6: Refocusing individual in Statement 7 is indeed thinking of refocusing attention from the "rinky-dink" to a different approach for students that will be more powerful. You should have underlined a more powerful and for students. The person is suggesting a far different innovation, and it is to benefit students, the clients, making it a 6.

Statement 8's individual is considering how to go about learning. You should have underlined this phrase because it suggests the person is interested in finding out about coaches, which indicates SoC 1: Informational.

This has been a long and productive session. Next time we will return to this activity briefly in order to segue into learning about the ways to collect Stages of Concern data and use them for professional development.

HANDOUT 4.1.A

Novice and Experienced Teachers

	Novice Teachers	*Experienced Teachers*
What about teaching are you most concerned with?		
What about teaching are you most satisfied with?		

HANDOUT 4.1.B

Typical Expressions of Concerns

	Stages of Concern	*Expressions of Concern*
IMPACT	Stage 6: Refocusing	I have some ideas about something that would work even better.
	Stage 5: Collaboration	I am concerned about relating what I am doing with what my coworkers are doing.
	Stage 4: Consequence	How is my use affecting clients?
TASK	Stage 3: Management	I seem to be spending all of my time getting materials ready.
SELF	Stage 2: Personal	How will using it affect me?
	Stage 1: Informational	I would like to know more about it.
UNRELATED	Stage 0: Unconcerned	I am not concerned about it.

Source: Figure 2.1, The Stages of Concern About an Innovation, from *Measuring Implementation in Schools: The Stages of Concern Questionnaire* (p. 8), by A. A. George, G. E. Hall, and S. M. Stiegelbauer, 2006, Austin, TX: SEDL. Copyright 2006 by SEDL.

HANDOUT 4.1.C

Practice Scoring Stages of Concern

Statements of Implementers	*Numeral and Name of SoC*
1. I'm going nuts trying to find time to create the right materials, while finding and organizing all the things necessary for this new curriculum.	
2. I'm new to this school, but I am studying and asking questions about the reading program.	
3. I worry that the students need a longer period of time for their end-of-year projects, but I think I can adjust this to help students have a more productive experience.	
4. You know, it's crazy, but I don't even know what the new math is all about.	
5. I think that we need to coordinate across our team's classrooms, so that all our kids are able to work collaboratively on their science projects.	
6. I don't know about the new history curriculum and whether I can handle this new approach. It is so different from what I am accustomed to.	
7. I'm thinking of chatting with our supervisor to suggest a more powerful technology approach for students in our school. What we brought in is such a rinky-dink strategy.	
8. I think we should consider how to go about learning if we will have coaches for the new social studies module.	

LEARNING MAP 4.2

Generating Responses to Concerns

Outcome

Learners will suggest assistance and support appropriate to each individual's Stage of Concern.

Assumption

One size does not fit all, and the assistance provided to implementers should be differentiated according to the individual's needs and concerns. In this way, implementers receive appropriate support so that they become more knowledgeable and effective with use of the innovation.

Suggested Time

60 minutes

Materials

1 copy for each participant of
  Handout 4.1.b, Typical Expressions of Concerns
  Handout 4.1.c, Practice Scoring Stages of Concern

Engaging in Learning

1. Request that participants access Handout 4.1.c, Practice Scoring Stages of Concern, and Handout 4.1.b, Typical Expressions of Concerns, that we used last session. The chart in Handout 4.1.b will help to remind you about the concerns definitions. Statements about concerns (such as these) provide the basis for understanding individuals' needs for assistance with an innovation in order to use it more productively.

2. We scored Statement 1 on Handout 4.1.c as which concern? Yes, 3: Management. Reflect a moment and imagine this person thinking about how to find time, and so on. If you are the change leader/facilitator or a collegial facilitator (colleague of the person) of implementation, what might you consider doing to support this person in planning and getting these management issues under control? Solicit responses from the group. Yes, sitting with this individual and generating a list of materials needed is a good start. Yes, encouraging the individual to invite teammates to a meeting where they can pool resources so that each one is not burdened is another good idea. Or suggest that this individual visit a colleague and ask for help in accessing and organizing materials.

   One of the most significant support actions we have ever seen was a principal who had shelves built into each teacher's classroom on which to organize and store materials.

There is typically not just one good idea, so keep notes; you might place these ideas in the Numeral and Name of SoC box on the right or make a running list of the stages and suggestions for responding to them—and create an inventory of these useful ideas for responding to each different Stage of Concern. Use your journal for recording, if that is a good place for your notes about Concerns.

3. Let's look at Statement 2. We determined last session that this statement reflected a person who has SoC 1: Informational concerns. What can we do to respond to this person's concerns? Of course, provide information—and there are myriad ways to share or provide information: telling the person what we know about the reading program, locating a clear but brief piece of text that describes the program, convening the grade-level team and asking members to describe the program, accessing videos showing teachers using the program, finding clips about the program online. The wise suggestion here is not to drown the person with information, but provide it in graduated steps so that each piece is integrated before adding more detail and richness.

4. Now, we have worked through two of the statements together. With a partner, discuss each of the statements and its rating, then discuss supportive actions that can be done to respond to each person's concern. Reach agreement, then write your suggestions in the Numeral and Name of SoC box or on the back of Handout 4.1.c. When you have completed this work, we will review it.

5. Provide a break, and on return, continue to discuss the possibilities for responding to each individual's concerns with action for assistance or support.

6. Let's look at Statement 3. How did we code that last time? Yes, it is a Stage 4: Consequence concern. Remember that this is an Impact concern. The person making this statement has conquered management and has moved to Consequence, where the speaker is wondering about the students and their needs. This is a frequent type of Consequence concern. This person needs help in clarifying his or her worry about the students, so any assistance that suggests ways to assess whether the students need more time or to clarify the status of the students relative to their projects would be useful. Other suggestions?

   For Statement 4, we marked this person as 0: Unconcerned, for this person knows nothing about new math. If this person is not due to be involved with new math, ignore the statement. But if the individual is expected to be engaged with new math, then the challenge is to push him or her to want information about it, that is, to move the individual to the next stage, Information. This may be done by pointing out the virtues of new math for students, citing the grade level or department's team in agreeing to use new math, or remind him or her (if all else fails) that this is the new district curriculum, and thus, we are all obligated to use it . . . and I (as the change leader of facilitator) will help you get started, bit by bit. The challenge is to push and to support simultaneously.

7. We coded Statement 5 as 5: Collaboration, since the teacher is suggesting that the team come together and treat the students as one whole so that students might collaborate with other students across all the classrooms. The purpose is to benefit students by permitting this larger collaboration, thus, it was scored as a 5: Collaboration. To support this endeavor, the change leader may need to give help in arranging alternative schedules or alternative space for the groupings. There should also be discussion and suggestions about how the teachers will assess the value of this collaborative endeavor for the students.
8. Personal concern is clearly expressed in Statement 6. This person requires one-on-one attention and help. The teacher is lacking in confidence, so promising help—in small pieces and in frequent applications—is very important. This person needs TLC, tender loving care: kind support but also gentle pushing. Let the person know that you will return very soon after each visit. This allows the teacher to know he or she will receive help, but also that in your visit you will expect to see action on the part of the teacher. This person needs private conversations, never in a large group.
9. And here is the 6: Refocusing individual in Statement 7, who has in mind, or is exploring, a very different program or practice to enhance or replace the current one. Many teams, schools, or districts would like to see everyone using the district-selected innovation for at least one cycle, then assessing its effectiveness (and possibly, efficiency), before making any large changes—which is what Stage 6 is about. The Stage 6 person might be promised voice at the end of the cycle, an opportunity for suggesting alternatives.

   If this is the policy of the school or district, then encourage this implementer to "ride with the herd," but promise keen attention at some identified point to this user. If this alternative is adopted, then this user might be a good candidate for facilitating others in its adoption and implementation.
10. Statement 8 we coded as 1: Information, for this person is interested in learning how coaches will be used for the new curriculum. The response should be fairly straightforward—the information is provided by a reliable source in a simple format, verbal or written.

Closure

We have engaged in the second part of a two-part activity: identifying individuals' concerns and then responding to them. It is a waste of time and effort to collect concerns data if they are not used.

In our next session, we will learn three means by which to collect concerns data and when each method will be most appropriate.

HANDOUT 4.1.B

Typical Expressions of Concerns

	Stages of Concern	*Expressions of Concern*
IMPACT	Stage 6: Refocusing	I have some ideas about something that would work even better.
	Stage 5: Collaboration	I am concerned about relating what I am doing with what my coworkers are doing.
	Stage 4: Consequence	How is my use affecting clients?
TASK	Stage 3: Management	I seem to be spending all of my time getting materials ready.
SELF	Stage 2: Personal	How will using it affect me?
	Stage 1: Informational	I would like to know more about it.
UNRELATED	Stage 0: Unconcerned	I am not concerned about it.

Source: Figure 2.1, The Stages of Concern About an Innovation, from *Measuring Implementation in Schools: The Stages of Concern Questionnaire* (p. 8), by A. A. George, G. E. Hall, and S. M. Stiegelbauer, 2006, Austin, TX: SEDL.

HANDOUT 4.1.C

Practice Scoring Stages of Concern

Statements of Implementers	*Numeral and Name of SoC*
1. I'm going nuts trying to find time to create the right materials, while finding and organizing all the things necessary for this new curriculum.	
2. I'm new to this school, but I am studying and asking questions about the reading program.	
3. I worry that the students need a longer period of time for their end-of-year projects, but I think I can adjust this to help students have a more productive experience.	
4. You know, it's crazy, but I don't even know what the new math is all about.	
5. I think that we need to coordinate across our team's classrooms, so that all our kids are able to work collaboratively on their science projects.	
6. I don't know about the new history curriculum and whether I can handle this new approach. It is so different from what I am accustomed to.	
7. I'm thinking of chatting with our supervisor to suggest a more powerful technology approach for students in our school. What we brought in is such a rinky-dink strategy.	
8. I think we should consider how to go about learning if we will have coaches for the new social studies module.	

LEARNING MAP 4.3

Collecting Concerns Data

Outcome

Learners will describe two methods for collecting Stages of Concern data and match the appropriate method to a specific purpose; learners will identify a third data collection method and its purpose.

Assumption

Concerns theory is applicable to a variety of circumstances. The particular method of collecting Concerns data is thoughtfully and purposefully related to the purpose for which it is being collected.

Suggested Time

90–120 minutes

Note: Three sessions might be devoted to this topic, one data collection process per session.

Materials

1 copy for each participant of

Handout 4.1.b, Typical Expressions of Concerns
Handout 4.3.a, Collecting Open-Ended Statements
Handout 4.3.aa, Scoring Open-Ended Statements
Handout 4.3.b, Scoring Informal Interviews
Handout 4.3.c, Referencing the 35-Item Questionnaire

Engaging in Learning

1. There are three SoC data collection processes: Open-Ended Statements, Informal Interviews, and the 35-item Questionnaire.

   We will examine the open-ended statements process first.

   As suggested by the label, this process focuses on obtaining statements from the implementer in response to an open-ended statement:

   "When you think about your role as a change leader or change facilitator, what are you concerned about? Do not say what you think others are concerned about but only what concerns you *now.* Please be frank and respond in complete sentences." Refer participants to Handout 4.3.a, Collecting Open-Ended Statements, and ask them to respond on the handout.

   Provide 5–10 minutes for participants to write.

   Refer the participants back to Handout 4.1.c, and explain to them that these statements are derived from asking the question above. Remind them how we scored those statements: reading the statement(s) for an overall sense of it; underlining words or phrases that suggest one of the Stages and placing the numeral for that Stage above the underlined portion to name it. When all words or phrases had been underlined and labeled, we determined from the marked and numbered clues and from viewing the SoC chart what the individual's concerns are.

Now ask participants to select a partner and underline and label their own statements, discussing with each other and coming to agreement on the labeling. Tour the two-person teams, providing assistance in their scoring. After their scoring, invite them to select one of their statements to read to the whole group, requesting the group score the statement. Applaud correct efforts and correct inappropriate scorings. Share with them that not all statements may contain a concern and will be extra verbiage and to ignore that. Solicit questions and needs for clarification.

Let us score another set of statements that were accessed just as we experienced, noting that the question will have asked for concerns about whatever the innovation may be in that setting (e.g., new math text adoption, middle school professional learning community, practices for teaching with technology tools). Refer to Handout 4.3.aa. With a partner, discuss each statement, marking clues, making a consensus decision about the labeling of the SoC, and placing the numeral and name of the concern in the blank at the end of each item.

Discuss their 4.3.aa responses:

1. Score this SoC 3, underlining organize all these materials, locate, and find the time.
2. Score this SoC 1, underlining need to find out, I am reading about it, and asking questions.
3. Score this SoC 2, underlining I will need to give up and will I be able to participate in decisions.
4. Score this SoC 5, underlining grade-level team came together, regrouped our students, to their social studies interests, they would learn more, and we could support them more.
5. Score this SoC 4, underlining my students need more practice and looking for extra stuff that will work.
6. Score this SoC 0, underlining have no idea of what [it] is.
7. Score this SoC 6, underlining students have mastered, ditch this, and think about using something like.

Solicit questions; turn to the whole group for answers, but correct them if necessary.

2. We have gained understanding and skills in soliciting open-ended statements and in scoring them.

3. Now, let's investigate the Informal Interview process.

   The Informal Interview process is much like the open-ended statements process, except that it is oral and not written. This informal chat or conversation is the least formal of the processes, and many times the implementer is not aware that concerns data are being solicited, noted, and deciphered. The chat will be most candid and productive if it occurs in a comfortable location, is conducted by a change leader or facilitator who is pleasant, calm, and interested in the implementer's responses (whether positive or possibly negative), and does not involve the pressure of time—although this little conversation rarely requires more than 5 minutes.

It may occur in the implementer's classroom or as the leader and implementer walk to the cafeteria or through the parking lot to their cars. It can be done over lunch or by some other prior arrangement. Again, it should not consume much time.

The goal is to engage the implementer in talking about how his or her work with the innovation is going, so that clues may be gained about concerns from the comments of the implementer. Here are a few possible commentary initiators:

How is it going with __________ (the innovation)?
What do you think of it?
How do you feel about it?
Any concerns that you have about it?
How does it affect you?

The facilitator does not use all these questions—they are possibilities to get the conversation started. As the implementer reports about his or her use and/or how he or she is feeling about it, the facilitator can, for clarity, pursue particular comments made by the implementer.

The clues that are gleaned are not recorded (on a clipboard or anything else) during this interaction, thus the change facilitator must remember what was heard and use it to analyze and score for concerns. We have seen change facilitators dash around the corner after such an informal interview to make quick notes, and score these notes, so as to identify the individual's concerns, on Handout 4.3.b, Scoring Informal Interviews, and to consider actions to be taken to support the individual.

4. We have learned to use the Open-Ended Statements process as well as the Informal Interview to gain data about an individual's concerns.

We now explore the 35-Item Questionnaire to gain data about concerns.

While the questionnaire may be used for facilitation purposes by experienced SoC users, it is most typically employed for research and evaluation—and is recommended for such. This use of the questionnaire is the topic of another book (George et al., 2006). The exploration of the questionnaire in this book is simply meant to introduce the reader to it and to give a smidgeon of information about it.

The questionnaire is constructed of 35 items; five statements are provided for each of the seven stages. As a result of the respondent marking the items, a profile can be drawn to illustrate the intensity of each of the seven stages. The peaks and valleys of the profile indicate which concerns are most intense (and which are lower) for an individual. The profiles may be superimposed on a single graph so that individuals or school profiles may be expressed simultaneously.

The questionnaire provides a quantitative score of the concerns, thus it is more precise—which is certainly an advantage. However, it does not provide the qualitative information that surrounds the open-ended and informal interview processes. The questionnaire expresses a keener score—*what* the concern is—but it does not suggest any possible *why* or *how* that is frequently included in the open-ended statements and informal interview.

An example of profiles generated by the questionnaire is shown on Handout 4.3.c, Referencing the 35-Item Questionnaire. In this example we see the profiles of four teachers, members of the fifth-grade team, who have responded to the questionnaire and whose scores have been plotted.

How would you interpret Martin Joe's profile? Yes, he has what we call the typical non-user profile: highest peak is on SoC 1: Informational, with the slope dipping continuously down. Since his highest and most intense SoC is Informational, what should be done to help Martin Joe resolve his Informational concerns? Of course, provide information, which can be done in a variety of ways: face to face, through written materials, by viewing a video clip, by visiting a classroom where the innovation is fully implemented.

Let's look at Susan's profile. What do you think of this? Susan is also expressing a non-user profile but it is significantly different from Martin Joe, in that Susan shows SoC 2: Personal concerns highest. It is important to notice this difference from Martin Joe, who is expressing SoC 1.

For Susan, short bits of information delivered in a personalized way, with easy-to-understand directions, is the approach. Express interest and warmth for Susan, but keep her feet to the fire with short assignments, always letting her know that you will return in a few days to check to see how things are going—and to add more to her knowledge and skills development. The "tail-up" on SoC 6 may suggest that Susan is thinking of an entirely new/different innovation and ditching the current one. Check this out, for the combination of SoC 2 and 6 suggests that immediate help is required for Susan, or she will make every effort to drop the current new practice.

Donna has an interesting profile, with her highest peak on SoC 4: Consequence. This means that Donna is questioning how well the innovation is influencing and impacting students. She is interested in assessing her work with students, and now is the time when tools or approaches for doing that are appropriate to introduce.

Last is Roger's profile. We see here the highest peak on SoC 5: Collaboration and the second highest on 4: Consequence. The purpose of the collaboration here is to design and deliver the innovation in ways likely to result in increased outcomes for students (as indicated by the second highest peak); therefore, SoC 5 is warranted. Ways to enable Roger and his colleagues to collaborate for kids' benefit should be explored: time to meet, a location for doing so, any materials or consultants that could be helpful.

In Conclusion

The Open-Ended Statement process for collecting data is an excellent choice when planning for professional learning sessions about the innovation. With these data, the planners can provide adult learning activities that are more on target for the needs of the implementers—as revealed by their SoC data. These written statements may contain relevant and useful qualitative information that the change facilitator may add to the SoC rating, giving a richer picture of the implementer's role and feelings about the innovation.

As already noted, the Informal Interview is the least formal approach for gathering data, and as often as not the implementer does not recognize that data are being collected. In addition, the facilitator's questions serve as a signal from the facilitator that there is interest, and importance, in the implementer's work. Simply arranging for a conversation and engaging in it suggests to the implementer that "perhaps I should give this my best effort." Because these are one-to-one inquiries, and are short in duration, they enable the facilitator to gain significant information quickly about the individual. Because these approaches target a single implementer at a time, they will likely not be used widely, but are excellent ways to tease out what the implementer is thinking, what he or she is concerned about, and can generate information for designing substantial assistance activities. This approach would be well used by coaches in their work with innovation.

The 35-Item Questionnaire is most fruitfully used for research and evaluation purposes when implementing innovations. If the change leader/facilitator explains in easy-to-understand terms why the questionnaire is being used, people can usually accept the instrument readily and respond to it. The questionnaire is, however, the most formal of the means by which to collect concerns data, so individuals can be put off by it. It is wise to administer the instrument to all implementers at the same time, so that they all hear the same explanation. Scoring the instrument can be done by hand or by computer program, but that is the topic of another day. As noted earlier, we mention this approach to collecting data, not because a facilitator is likely to use it for facilitating use of an innovation, but because it exists—thus, it is an item for the participants' knowledge base, with skills development done in another project.

HANDOUT 4.1.B

Typical Expressions of Concerns

	Stages of Concern	*Expressions of Concern*
IMPACT	Stage 6: Refocusing	I have some ideas about something that would work even better.
	Stage 5: Collaboration	I am concerned about relating what I am doing with what my coworkers are doing.
	Stage 4: Consequence	How is my use affecting clients?
TASK	Stage 3: Management	I seem to be spending all of my time getting materials ready.
SELF	Stage 2: Personal	How will using it affect me?
	Stage 1: Informational	I would like to know more about it.
UNRELATED	Stage 0: Unconcerned	I am not concerned about it.

Source: Figure 2.1, The Stages of Concern About an Innovation, from *Measuring Implementation in Schools: The Stages of Concern Questionnaire* (p. 8), by A. A. George, G. E. Hall, and S. M. Stiegelbauer, 2006, Austin, TX: SEDL. Copyright 2006 by SEDL.

HANDOUT 4.3.A

Collecting Open-Ended Statements

When you think about your role as a change leader or facilitator, what are you concerned about?

Do not say what you think others are concerned about, but only what concerns you *now*.

HANDOUT 4.3.AA

Scoring Open-Ended Statements

1. I feel the need to organize all these materials so that I can locate and use them, and to find the time to do that. I am swamped. ____________________________ (Numeral & Name of SoC)

2. I just came on board to this school, and I need to find out what this new program is all about, so I am reading about it and asking questions of my colleagues. ______________________________ (Numeral & Name of SoC)

3. I am concerned that I will need to give up my professional role in this new program. Will I be able to participate in decisions? ____________________________________ (Numeral & Name of SoC)

4. I think if our grade-level team came together and regrouped our students according to their social studies interests, they would learn more and we could support them more effectively. ____________________________________ (Numeral & Name of SoC)

5. I am thinking that my students need more practice materials for their Spanish language development. I am thinking of looking for extra stuff that will work. _____________________________________ (Numeral & Name of SoC)

6. I have no idea of what this new elementary school calculus is. _____________________________________ (Numeral & Name of SoC)

7. My students have mastered the mechanics of seventh-grade mathematics already. I think we should ditch this and think about using something like a constructivist approach that will give them a chance at real problem solving. _____________________________________ (Numeral & Name of SoC)

HANDOUT 4.3.B

Scoring Informal Interviews

	Implementer (name)	*Implementer (name)*
What prompted the individual to comment on concerns?		
What is the individual's SoC?		
What is your evidence?		
What action will you take to assist or support this individual?		

HANDOUT 4.3.C

Referencing the 35-Item Questionnaire

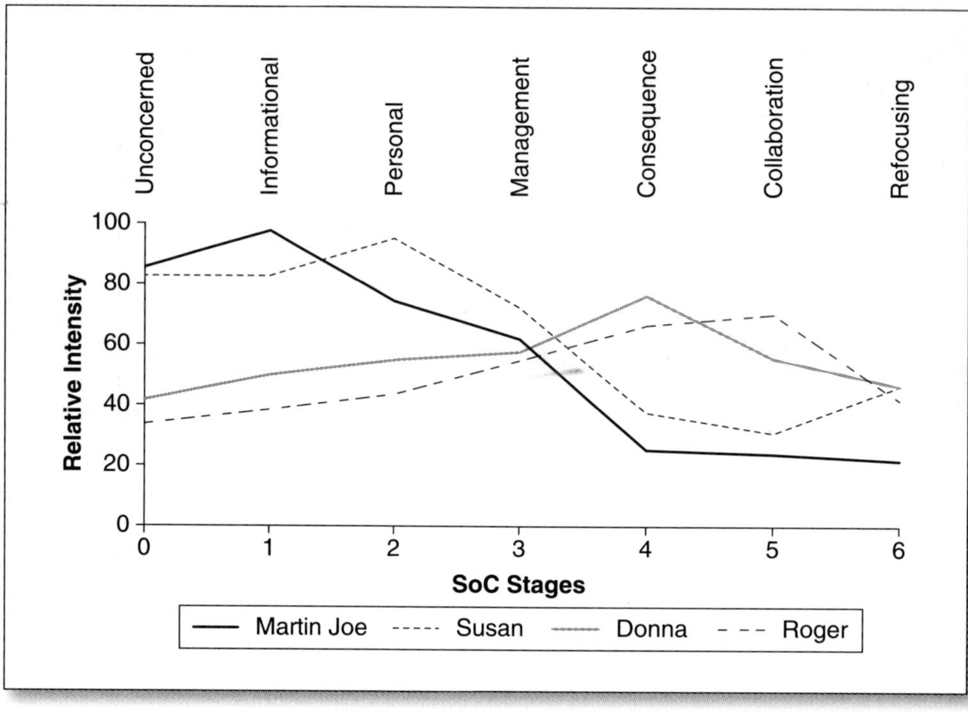

5

Levels of Use

Using Innovations

EVERYWHERE MEETS LEVELS OF USE

Bob Yurself's team now understood that there are different kinds of worries that novice and veteran teachers experience during a change. By using the seven indicators that describe staff concerns, change facilitators can be much more effective in supporting people during implementation. And understanding how to apply the three different ways to collect Stages of Concern data, staff can feel that their concerns have been heard, and facilitators, including collegial facilitators, can better assist them in support that may be needed in implementing the change. The collection of these data also assists in planning for the right kind of professional learning staff may find valuable in making sense of the change. Stages of Concern will forever alter how EveryWhere guides an innovation forward in the future.

Bob has consistently encouraged all staff members to become involved with the concerns data at their grade level, or subject area, and to give careful thought to how they might support their colleagues. This collegial facilitation would expand the pool of persons offering support and assistance, and further, it would provide the opportunity for the staff to become personally involved in and committed to the change effort. Thus, Bob led the district team in developing skills for studying the concerns data, how to share it with teachers, and how to challenge teachers to suggest help for each of the Stages of Concern represented in the data. The district team would then teach these skills to their schools so that collegial facilitation would begin to develop across the district.

Bob was seeing another example of how change and gardening are so similar. Just as a gardener can water the soil when there isn't enough rain or weed the garden so more of the nutrients get to the plant, Bob can also do that same kind of tending when implementing change. As a change facilitator he and his team can use the Concerns-Based Adoption Model (CBAM) methods like Stages of Concern (SoC) and Levels of Use (LoU) to guide an innovation forward by being aware of and responsive to those directly impacted by the change. Just as Bob monitored weather conditions to determine what his garden needed to sustain it, he (and his team) can now attend to staff members' concerns as they experience a change in order to better sustain the innovation.

Bob and his team are now ready for the last important chapter of their own learning: Levels of Use. While Stages of Concern focuses on the affective side of change—feelings, reactions, emotions, or attitudes—Levels of Use directs attention to the behavioral dimension, or the approach the implementer is taking in relation to the innovation. Understanding the Levels of Use concept and the critical data that result from measuring participants' LoU provides another basis for determining appropriate kinds of assistance and support that staff may need. LoU information will be helpful in showing how EveryWhere staff are making sense of the math changes in their daily work.

As he did with the SoC concept and its data, Bob intends to encourage implementers to be sensitive to their colleagues' LoU status through observation and through reviewing and interpreting the LoU data that they and their facilitators collect. Bob knows that every implementer will not be interested in adding Levels of Use to his or her repertoire, nor be skillful in this activity, thus he will not push it. But he will support all whom he can persuade to be involved. For in doing so, he feels that he will have shared with them a measure of the power, authority, and decision making that typically resides only in "top of the heap" leaders. He maintains this position in order to increase shared and collaborative leadership for implementation. Of course, Bob will remain accountable for the success of the change projects, thus he will continuously observe all the "players."

Innovation Configurations is the tool that articulates for implementers exactly what the new "thing" or innovation is. Quite differently, SoC and LoU inform us about the individual user or implementer: SoC reveals the reactions, attitudes, or affect of the individual about the change; LoU portrays the behaviors, or approaches, the individual is taking with the change. Few change leaders or facilitators are able to use both SoC and LoU to help guide a change effort (primarily because of the time required to do so). Thus, many use one or the other. Many change leaders opt to use SoC, possibly because it has been more visible in the literature. Others prefer LoU because it is behavioral and is thus observable and easier to gain information and insight from.

In any case, you as the reader and learner have the choice. And to that end, we now turn to exploring and understanding Levels of Use.

LEARNING MAP 5.1

Articulating Behaviors Associated With the Use of Innovations

Outcome

Learners will describe eight specific behaviors associated with individuals' learning to use innovations.

Assumption

Seldom do individuals move in one step from not being able to use innovations to using them in a quality way. The concept of Levels of Use identifies eight research-based levels through which individuals move, from no use at all to various descriptions of behaviors in their use, as they implement an innovation. The implementation facilitator uses these identifications to create the most appropriate help or assistance to support individuals in continuous movement to appropriate levels.

Suggested Time

60 minutes

Materials

1 copy for each participant of
Handout 5.1, Typical Behaviors, Levels of Use

1 set of index cards (8), with one LoU title on each card

Engaging in Learning

1. While Stages of Concern describes the affective side of change—the feelings, reactions, emotions, or attitudes toward change—Levels of Use directs our attention to the behavioral dimension, or the approach that the implementer is taking in relation to the innovation. Handout 5.1, Typical Behaviors, Levels of Use, gives us a crisp description of these behaviors. Please refer to this handout as we discuss each of these levels.
2. A broad cut at these levels results in two large categories: users and nonusers. Reviewing the levels from the bottom up, we begin by giving attention to non-users. LoU 0, I, and II constitute the non-user categories. The first of these refers to any individual who is taking no action whatsoever related to the new practice or program and is labeled LoU 0: Non-use. This nonuser may be completely unaware of the existence of the innovation or, if so, is giving it no attention.
3. On the other hand, the LoU I: Orientation person is aware of the new program or practice and is taking action to gain information about it. He or she may be asking others about it, studying materials, attending workshops, and so on. The key here is actively looking for and gaining information, or making plans to do so.

4. At LoU II: Preparation, the non-user is preparing to move into using the innovation. We have learned that not only is preparation underway, but the individual has identified a specific time at which use will begin. Such preparation actions may include collecting materials and equipment or interfacing with others to gain information and understanding about using the innovation. Different from LoU I, where the information is sought to obtain knowledge of it, at LoU II the purpose of the information is to aid the individual to begin to use the innovation.

   Please note that the numerals used for LoU are Roman, whereas the numerals for SoC are Arabic—a small means to differentiate between SoC and LoU.

5. The remainder of the levels indicate Users, and the first of these is LoU III: Mechanical. Individuals at this level are seen as lacking order and organization, being unsure of what to expect next, and having difficulty handling time and materials. The time factor is typically one that is regularly bemoaned by the new user. The focus of attention on such factors stimulates the user to try multiple ways or changes to organize for use, with typical lack of immediate success. As at all levels, support and assistance here are critical.

6. As a result of help in resolving the LoU III: Mechanical factors, the individual may move to LoU IVA: Routine. This means that the user has found stability and has discontinued changes since he or she has found a way to make the innovation work for him or her. The Routine user typically expresses satisfaction with his or her use of the innovation. The response here is to investigate exactly *what* routine looks like. Using the IC map of the innovation can help to pin-point what the user is doing. If what the individual is doing is acceptable, then confirming and congratulating the user is in order. If, on the other hand, the person's placement on the IC map is on the poor end of the continua of the map components, then taking this into account should be done for creating help or assistance.

7. Unlike the IVA user, the LoU IVB: Refinement user is asking himself or herself how the new practice is working for clients—client benefit is the focus; the user has turned away from himself or herself. Assessing is a typical behavior, and changes to increase client outcomes are seen or are the target of the activity. This individual would appreciate ideas about how to assess the change and/or how it is being used.

8. LoU V: Integration labels the user who is making intentional efforts to work with other users and to coordinate how the innovation is being used; importantly, the purpose of this coordination is to benefit clients. If the intention is to reduce activities to coordinate time and materials, this refers to management; this would indicate the user at LoU III: Mechanical, and not LoU V: Integration. Thus, it is important to note the purpose for which integration is desired. To support authentic integration efforts, assisting the users with scheduling in order to meet, and obtaining resources and consultants (possibly) for their coordination work would be a large boon to them.

9. Very few users reach LoU VI: Renewal. This level is used to name those who are thinking of or seeking alternatives to the innovation under implementation, or considering major modifications or changes in the current innovation. In many change efforts, this individual is discouraged from such changes, until such time as the innovation has experienced a cycle of use—at which time, the LoU VI individual may be invited to assist or lead an effort in revising the innovation or in selecting or designing a replacement innovation.

10. Pause and solicit participants' questions about the levels and the rationale for rating individuals for each level.

11. Distribute the eight index cards to single participants, pairs, or triads (depending on the number in the group, making sure that all of the eight cards are distributed), asking each card holder to review the descriptions of the eight Levels of Use and directing him or her to create a description of a person who would be characterized at the LoU on the card. The card holder(s) will write the description on the back of the card, then share this description with the large group, and invite participants to report which LoU is being described. When a response is given, invite others to report if and why the rating is correct. If responses are not correct, explain and redirect thinking to the proper definitions of the levels.

HANDOUT 5.1

Typical Behaviors, Levels of Use

	Levels of Use	*Behaviors Associated With LoU*
USER	Level VI: Renewal	Explores major modifications or alternatives to current innovation
	Level V: Integration	Coordinates innovation with other users for increased client impact
	Level IVB: Refinement	Makes changes to increase client outcomes, based on assessment
	Level IVA: Routine	Makes few or no changes to an established pattern of use
	Level III: Mechanical	Makes changes to better organize use
NON-USER	Level II: Preparation	Prepares to begin use of the innovation
	Level I: Orientation	Seeks information about the innovation
	Level 0: Non-use	Shows no interest in the innovation; takes no action

Source: Table 2.1, Levels of Use of the Innovation, from *Measuring Implementation in Schools: Levels of Use* (p. 5), by G. E. Hall, D. J. Dirksen, and A. A. George, 2006, Austin, TX: SEDL.

LEARNING MAP 5.2

Identifying the Level of Use of Individuals

Outcome

Learners will identify an individual's Level of Use and suggest appropriate responses to the person.

Assumption

In addition to the behaviors chart, a second tool to help determine the LoU of an individual is introduced in this session and practice provided in identifying LoU. Creating responses to the identified LoU is also given attention, for simply knowing the LoU is time-wasting unless it is used productively.

Suggested Time

60 minutes

Materials

1 copy for each participant of
Handout 5.1, Typical Behaviors, Levels of Use
Handout 5.2.a, Decision Points of LoU
Handout 5.2.b, Identifying an Individual's LoU

Engaging in Learning

1. We will practice again identifying the Level of Use of an individual who is involved in the change process. Please access your copy of Handout 5.1, the LoU chart. Review it, and develop a mental image in your head that shows the behavior of a person at each level. Invite participants to share one of their mental images, and solicit others to identify the LoU. Compliment correct mental images and identifications, and correct faulty or incorrect items.
2. Distribute Handout 5.2.a, Decision Points of LoU. The Decision Points note in a very crisp, clear way exactly when a person moves to a Level of Use. For instance, beginning at the bottom, we see the LoU 0: Non-use person, who is taking no action, doing nothing. You will undoubtedly notice the strong parallel to the LoU behaviors chart we studied last session. When the individual takes action to learn about the innovation, LoU I: Orientation is established. Place your copy of Handout 5.1 alongside Handout 5.2.a, and notice the closely stated definitions of each level on each page. Review each level, and share any questions that you have. Respond to questions.

3. Using both Handouts 5.1 and 5.2.a, let us practice identifying the LoU of an individual on Handout 5.2.b, Identifying an Individual's LoU. Let's do the first one together. Read the statement aloud together in unison.

   What words or phrases give you clues about this individual's LoU?

   Yes, that's correct. Underline <u>units were used and completely tested</u> and <u>worked well</u>; they have found a way to make them work. They will change nothing and <u>use them in the same way this year</u>. In the blank, write IVA: Routine. Which Decision Point confirms this rating? Yes, D-1. Write D-1 in the right margin. Any questions about that? Needs for clarification?

4. With a partner, discuss and score the next one, Statement 2. Did you mark <u>investigate . . . considering using</u> and <u>wants to know all about it</u>? The LoU is I: Orientation, because the teacher is looking for information. What is the Decision Point? Yes, it's A; place that code in the margin alongside Item 2.

   Now, with your partner, read, discuss, underline the clues and come to agreement on the scoring. Place the LoU numeral and name on the blank, and the Decision Point in the right margin for the remainder of the statements.

5. Now, let's check your work. I will provide the correct responses and the rationale for them. For Statement 3, you should have underlined <u>added a major project</u> and <u>completely changed the tone and tempo of the course</u>. This is scored LoU VI: Renewal because a major modification/addition has been made to the program. Decision Point F should be designated.

   For Statement 4, you should have underlined <u>initiate learning teams next semester</u> and <u>fifth-grade teachers will begin preparing</u>, making this LoU II: Preparation. Decision Point B should be assigned on this statement. Questions? Comments?

   For Statement 5, you should have underlined <u>figure out how to arrange and store</u>, <u>not working</u>, and <u>tried several ways to do it</u>, which indicates LoU III: Mechanical. The Decision Point should be C.

   For Statement 6, you should have underlined <u>two chemistry teachers are joining forces</u> and <u>so that the students can select . . . better fit their learning styles</u>, indicating LoU V: Integration, since the teachers are working together to serve their students more effectively. This item is Decision Point E.

   Invite questions and comments. Ask for needs for clarification.

   Next session we will learn how to collect LoU data. That comes, of course, before scoring it.

HANDOUT 5.1

Typical Behaviors, Levels of Use

	Levels of Use	*Behaviors Associated With the LoU*
USER	Level VI: Renewal	Explores major modifications or alternatives to current innovation
	Level V: Integration	Coordinates innovation with other users for increased client impact
	Level IVB: Refinement	Makes changes to increase client outcomes, based on assessment
	Level IVA: Routine	Makes few or no changes to an established pattern of use
	Level III: Mechanical	Makes changes to better organize use
NON-USER	Level II: Preparation	Prepares to begin use of the innovation
	Level I: Orientation	Seeks information about the innovation
	Level 0: Non-use	Shows no interest in the innovation; takes no action

Source: Table 2.1, Levels of Use of the Innovation, from *Measuring Implementation in Schools: Levels of Use* (p. 5), by G. E. Hall, D. J. Dirksen, and A. A. George, 2006, Austin, TX: SEDL. Copyright © 2006 by SEDL.

HANDOUT 5.2.A

Decision Points of LoU

The individual . . .

Decision Point F: Explores alternatives to or major modifications of the current innovation > VI: Renewal

Decision Point E: Makes changes in innovation use in coordination with colleagues, for client benefit > V: Integration

Decision Point D-2: Changes innovation use based on assessment in order to increase results for clients > IVB: Refinement

Decision Point D-1: Establishes a routine pattern of use and is making no changes > IVA: Routine

Decision Point C: Uses and/or makes changes, if any, based on his/her needs > III: Mechanical

Decision Point B: Determines to prepare and to use the innovation by identifying a time to begin > II: Preparation

Decision Point A: Acts to learn and to gain more detailed information about the innovation > I: Orientation

No action > 0: Non-use

Source: Table 2.2, Levels of Use of the Innovation with Decision Points, from *Measuring Implementation in Schools: Levels of Use* (p. 7), by G. E. Hall, D. J. Dirksen, and A. A. George, 2006, Austin, TX: SEDL. Copyright © 2006 by SEDL.

HANDOUT 5.2.B

Identifying an Individual's LoU

Read each statement, underline clues that suggest the LoU, and put the numeral and name of the level in the blank.

1. The materials and activities for the critical thinking units were used and completely tested last year. They worked well, and we will use them in the same way this year.

   ____

2. The algebra teacher is attending a conference to investigate inquiry learning, which he is considering using with his seventh-grade class. He wants to know all about it before ordering materials.

   ____

3. Two months ago the health instructor added a major project to the first aid course for students, which took them to the county hospital for first-hand experiences. It has completely changed the tone and tempo of the course.

   ____

4. The elementary schools' language arts coordinator announced that the schools would initiate learning teams next semester, so our fifth-grade teachers will begin preparing to determine their learning needs.

   ____

5. Our science department head has worked with us to figure out how to arrange and store our equipment, but it is not working, although we have tried several ways to do it. We are weary of all the stuff getting in the way and not being able to find what we need. Our principal said that he would build some shelves for us.

   ____

6. Have you heard? The two chemistry teachers are joining forces to regroup students across their two classes so that the students can select technology tools and lessons online that better fit their learning styles.

   ____

LEARNING MAP 5.3

Collecting Levels-of-Use Data

Outcome

Learners will conduct informal LoU interviews to collect data about implementers and identify the individual's Level of Use.

Assumption

To identify the implementer's LoU related to a specific innovation, an interview is used. There are two forms of the interview, each serving a different purpose: (1) a brief, informal conversation between a change leader/facilitator and the implementer to ascertain the implementer's LoU and to provide assistance and support in the process of implementation and (2) a formal, in-depth interview focused on multiple factors, which results in an evaluation-based identification of the implementer on multiple dimensions. This second interview is used for research and evaluation, and it is conducted by a person who has had 3 days of in-depth training and development in the skills needed to do so.

This Learning Map focuses on the first interview process since the purpose of this book is facilitation of implementation and not to engage in research or evaluation. That is the topic of another book (Hall et al., 2006).

Suggested Time

60 minutes

Materials

1 copy for each participant of

Handout 5.1, Typical Behaviors, Levels of Use
Handout 5.3.a, The Branching Approach to LoU Identification
Handout 5.3.b, Identifying LoU From the Branching Interview

Engaging in Learning

1. Unlike Stages of Concern, which has three discrete methods for identifying an individual's SoC, Levels of Use has one process—an interview. As noted in the Assumptions of this Learning Map, there are two variations to the interview: a brief, casual version, designed to obtain information for responding to the individual's current LoU and supporting movement to higher levels, and a more formal process used for research and evaluation purposes. We will engage in the more informal method, using what has become known as the Branching Interview, or Approach.

2. We will work on developing our skills in interviewing to gain data about the implementer so as to score the LoU and provide appropriate support. Be sure to have Handout 5.1, our Levels of Use behaviors chart, for

handy reference. Also note our new Handout 5.3.a, The Branching Approach to LoU Identification.

Notice that there are two branches: left and right. They are differentiated by the question at the top: Are you using the innovation? This will remind you of the LoU chart we have been using, which separates non-users from users. Non-users are noted in the left column of the LoU Branching Approach.

The first question, noted above, separates the non-users from the users. Thus, the non-user response to this initial question is "no," and the interviewer moves to questions on the left branch. The interviewer follows the questions from top to bottom, taking options from the interviewee's response, for the next question.

Through a series of eliminations, the specific rating of the individual is reached; the left branch identifies the LoU of a person who is not yet using the innovation as an LoU II, I, or 0. These classifications are parallel to those on the LoU chart, which is also divided into non-users and users.

If the answer to the first question—Are you using the innovation—is "yes," then the interviewer uses the questions on the right branch, following from top to bottom, using the questions to eliminate possible LoU ratings and to move to the next question. Please study the right branch to see how it permits the interviewer to find the precise LoU of the individual.

Questions?

3. Now let us use Handout 5.3.b to understand this process better. Please read Statement 1 and imagine an interviewer having asked this person, "Are you using the innovation?" The interviewee reports, "Yes," and responds to the next question on the right side of the Branching Approach: "What kinds of changes are you making in your use of the innovation?" The response indicates no changes are being made. Thus, the rating is LoU IVA: Routine. This rating should be written in the blank under Statement 1.

4. With a partner, please do Statement 2, and we will share our answers.

   The response to the initial question, "Are you using . . . ?" is a "no." The next question on the left branch is then asked: "Have you decided to use it and set a date to begin use?" The interviewee responds clearly that use will begin in 6 weeks and preparation is underway for that. The rating is LoU II: Preparation and should be so noted in the blank space.

   Now, with your partner, continue and complete the final three items, using the same procedure, with the Branching questions.

   Now that you have completed those tasks, let us check your responses.

5. Checking your work, look at Statement 3. The "Are you using" question elicits a "yes" response; the following question, "What kinds of changes are you making . . . ?" gains the explanation that "we . . . are not messing with it." No changes, thus the rating is LoU IVA: Routine.

For Statement 4, the answer to "Are you using" is yes, and the "kinds of changes" question produces an answer that includes "trying to figure out . . . schedule . . . distribute all the materials . . . organizing the kids." The rating is LoU III: Mechanical.

Statement 5 provides a "yes" to the "Are you using" question, and the "kinds of changes" question leads to the possibilities of impact-oriented changes. The "Are you coordinating your work" question gets a "yes" answer. So the rating for this interviewee is LoU V: Integration, certainly, and if the internship is deemed sufficiently major or large, then LoU VI: Renewal is also indicated.

Ask for questions or needs for explanations.

Congratulate participants on their new knowledge and skills.

HANDOUT 5.1

Typical Behaviors, Levels of Use

	Levels of Use	*Behaviors Associated With the LoU*
USER	Level VI: Renewal	Explores major modifications or alternatives to current innovation
	Level V: Integration	Coordinates innovation with other users for increased client impact
	Level IVB: Refinement	Makes changes to increase client outcomes, based on assessment
	Level IVA: Routine	Makes few or no changes to an established pattern of use
	Level III: Mechanical	Makes changes to better organize use
NON-USER	Level II: Preparation	Prepares to begin use of the innovation
	Level I: Orientation	Seeks information about the innovation
	Level 0: Non-use	Shows no interest in the innovation; takes no action

Source: Table 2.1, Levels of Use of the Innovation, from *Measuring Implementation in Schools: Levels of Use* (p. 5), by G. E. Hall, D. J. Dirksen, and A. A. George, 2006, Austin, TX: SEDL.

HANDOUT 5.3.A

The Branching Approach to LoU Identification

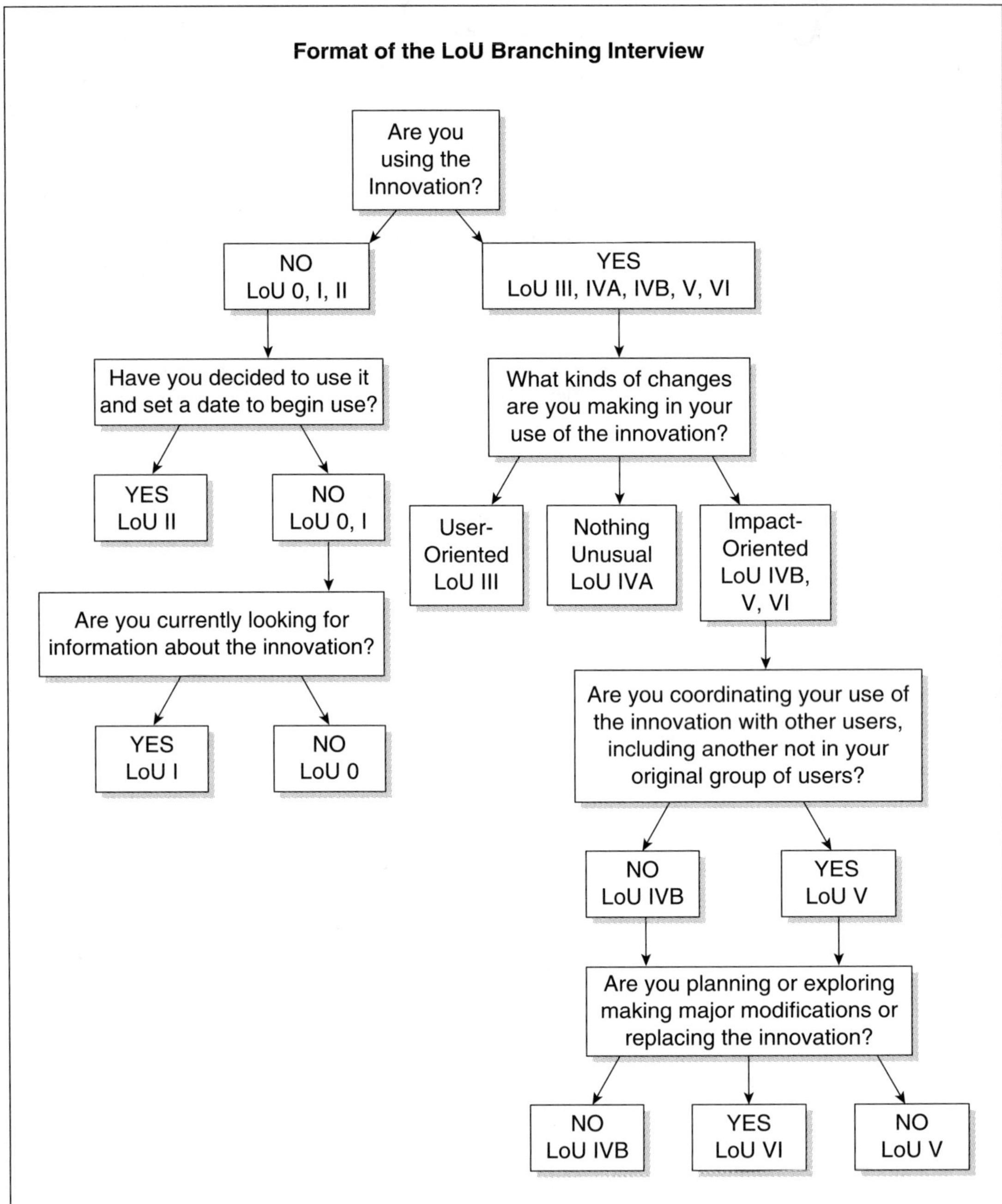

Source: Figure 3.1, Branching Chart, from *Measuring Implementation in Schools: Levels of Use* (p. 18), by G. E. Hall, D. J. Dirksen, and A. A. George, 2006, Austin, TX: SEDL. Copyright © 2006 by SEDL.

HANDOUT 5.3.B

Identifying LoU From the Branching Interview

Read each statement. Notice how the individual answers the question "Are you using the innovation?" and the additional information about his or her use. Place the numeral and name of the LoU on the blank.

1. Yes, and I am so relieved. I finally determined how to input the new problem-solving program into other class work, and now that I know what I am doing, I shall stay in that pattern and not change it.

   ____

2. Oh, dear, no. I have no idea yet what I will be doing with the technology innovation, but since we will start it in 6 weeks, I am getting information and the equipment together.

   ____

3. Goodness, yes. Our team shares what we each are doing with the critical thinking process. Thank goodness, we have it in stride and are not messing with it. We just share our triumphs and tribulations about it.

   ____

4. Yikes, yes; but I just wish I had the new online program in good shape. I have messed around trying to figure out when to schedule it, and how to distribute all the materials, in addition to organizing the kids and the computers.

   ____

5. Yes! It is so exciting! My teammates and I are collaborating to redesign our community service project so that the students gain insights into the purposes of all our community agencies' service facilities and how the agencies may be available for internships so that they learn more about engagement in community service.

   ____

EveryWhere's Journey Comes to an End, or Does It?

It is now the end of the third year since Bob Yurself was assigned as the district's change coach. It has been three extraordinary years of learning for Bob, his change leadership team, and staff members across the district. Happily, there is now plenty of evidence that EveryWhere District has turned the corner for improving mathematics learning.

The journey of change through which Bob and his team guided the district was viewed as the best professional learning experience in which the district ever participated. Clearly, learning has been happening on so many different levels. Staff have been professionally learning in ways they never had before, and subsequently, students are performing more successfully in math.

For example, teachers across the district in all three levels of the schools—elementary, middle, and high schools—regularly monitor student learning data and meet together in professional learning communities to discuss what the data mean and what they, the educators, need to learn in order to make improvements. In each of the schools there is the articulated belief that everyone's perspective and understanding are necessary for making a difference in student learning. This means that staff, over the 3 years, have invested time to hear diverse perspectives and are open to new ways of taking action. The staff have modeled with each other the most important human traits for improving schools in the 21st century: resiliency and adaptability, along with responsibility and accountability. EveryWhere staff are eager to take on the next change challenge and work together to generate the collective impact they now see in math.

Something was also changing with district leadership. They no longer clung to broadcasting mandates or prescribing change via a top-down approach. EveryWhere leaders act as social architects for guiding school improvement. Just as an architect designs physical space, district leaders' job is to create a social environment where people can identify and act on what matters to them, in the best interests of students. It is from this conversational space that authentic commitment and accountability for change happen best. As change leaders, EveryWhere administrators and teacher leaders see themselves attending to the social fabric of school life and designing change through interactions, conversations, and questions that invite personal and social investment.

Another significant shift across the district relates to communication and reflection. How staff members talk to one another and engage each other is a lot different from the past. In each of the buildings there is much more listening and asking questions to support a reflective culture. There is also a lot less telling and being directive. It is now understood that learning occurs through the social interactions that staff have with each other. The more respectful and honoring these interactions are, the more potential exists for increasing learning and exploring new behaviors. The carefully guided implementation of the mathematics changes, grounded in the use of the Concerns-Based Adoption Model (CBAM) concepts and approaches, created a culture of inquiry and improvement that incorporated the skills one might see in good coaching. Communication in EveryWhere is viewed as the social lubricant that makes learning and change sustainable and successful.

Bob Yurself came away from this experience with very important understandings about guiding change. He made a note to himself: "The knowledge base that has resulted from rigorous research on change and on change leadership has given school and district educators the concepts, tools, techniques, and approaches that can increase the probability of success in change projects—the research-based constructs in CBAM are key elements to be employed for that success." Bob realized that the learning and application of the CBAM tools have a flow to them that made implementation of a change highly successful. He wanted to remember that flow so he could replicate his change success in the future. As Bob reflected, he could see a guiding change series of spiraling steps appear.

IMPLEMENTING CHANGE THROUGH LEARNING

The following is Bob's set of ideas and activities. You can take it as your own or add to it in order to guide future planning of changes that will make a difference with staff and students in your school or district.

❒ Identify and communicate appropriate data to validate this change—It is important to remember that *one* source of data indicating a problem doesn't mean staff will quickly run off and change what they were currently doing in order to shift to something different. Staff members need to see multiple sources of data and especially the kinds of data that are valid from their point of view.

❒ Create a sense of honest urgency for change—There are two ways you can create urgency in change. You can do it through fear or you can do it through hope. Our experience has shown us that when you align a change initiative with people's values and their hope for a more positive and engaging future, the urgency that is evoked taps into the positive side of the human spirit.

❒ Explore the *why* of the change through dialogue before getting to the *what* or *how*—It is important to build a common understanding for the need to change. This can happen best not only through the examination of data, but also with open dialogue about the data. Invite collaboration by requesting different opinions, giving everyone ample opportunity to share perspectives about the data and what they think are the surrounding issues relating to the change.

❒ Identify critical stakeholders—A stakeholder represents anyone who will be affected by the change either directly or indirectly. If you give careful attention to those who represent a stakeholder interest in the change event you are exploring, you will have a greater chance of making the change sustainable over time. Remember, a change process is only successful in the long run if the stakeholders truly adopt and sustain the change.

❒ Check for staff readiness of the change before starting implementation—Use the Readiness for Change checklist (Handout 2.4) to assess readiness.

❒ Structure the social conditions so people can safely interact around the change; create psychological safety so the right conversations can happen.

❒ Create a 3- to 5-year plan for a change effort focused on the six strategies in order to cross the implementation bridge. (See Handout 2.2.b on constructing a skeletal plan for crossing the implementation bridge.)

❒ Provide opportunities for learning communities to emerge around the change. (Refer to the article "Professional Learning Communities: What Are They and Why Are They Important?" in Learning Map 3.3).

❒ Create a shared vision for the change by developing the Innovation Configuration (IC) map. (Refer to Learning Maps 3.2 through 3.6.)

❒ Review and critique the IC map in order to clarify components and ensure relevant variations—Remember, the IC map is not a rubric; it simply and clearly tries to describe the variations of an innovation as it is being implemented. (Refer to Learning Map 3.5.)

❒ Field test and revise the IC map—Initiate conversations with staff on the IC map using open-ended questions, such as the following. (Refer to Learning Map 3.6.)

Please tell me, if you will, about the new . . . (whatever your IC map focuses on)

What would I see you doing?

If I were the proverbial fly on the wall, what would I observe?

❒ Create a plan for sharing the IC map with implementers—Teach implementers *what* the map is, *why* we have it, and *how* to use it. This will ensure that the map is well used and results in impact on educators and, subsequently, on students. (Refer to Learning Map 3.7.)

❒ Use the IC map to initiate the development of an Implementation Plan for the change. (Refer to Learning Map 3.8.)—You might also invite semimonthly dialogues and discussions around the IC map for how the change is moving forward and what staff are or are not learning.

❒ Review with the change team the concept of Stages of Concern (SoC)—The goal of SoC is to determine how the implementer feels about the innovation. These concerns represent an individual's feelings, reactions, or attitudes (affect) about the change. (Refer to Learning Map 4.1.)

❒ Generate and collect SoC responses from implementers—Remember, one size does not fit all, and the assistance provided to implementers should be differentiated according to the individual's needs and concerns. Determine staff members' Stage of Concern for this innovation and the appropriate support and assistance. (Refer to Learning Maps 4.2 and 4.3.)

❒ Review with the change team the concept of Levels of Use (LoU)—Levels of Use are eight research-based levels through which individuals move, from no use at all to various descriptions of behaviors, as they implement an innovation. The implementation facilitator uses these identifications to create the most appropriate help or assistance to support individuals in continuous movement to appropriate levels. (Refer to Learning Map 5.1.)

❒ Conduct informal LoU interviews to collect data, and identify the individual's Level of Use—Conduct a brief, informal conversation with the implementer to ascertain the implementer's LoU and to provide assistance and support for the process of implementation. (Refer to Learning Maps 5.2 and 5.3.)

❒ Watch for indicators of implementation success as implementers move across the Implementation Bridge (SoC, LoU, and IC)—They use the strategies to move across the bridge and use the three tools to know how to construct or modify the strategy to best support the individual.

❒ Celebrate milestones of success along the way as implementers move across the Implementation Bridge.

❒ Consider the following guiding beliefs during a change:

- All change is based on learning, and improvement is based on change.
- Implementing a change has greater success when it is guided through social interaction.
- Individuals have to change before the school can change.
- Change has an effect on the emotional and behavioral dimensions of humans.
- People will more readily choose to change when they foresee how an innovation will enhance their work.
- A role of change leaders is to facilitate the conversations that invite others in owning the desired change.

❒ Think of resistance as your friend, rather than the enemy of change—How might you use those who are resisting to gain insight and understanding around the challenges of implementing the change? What hasn't been communicated around the change that is igniting resistance? How might Stages of Concern help you in understanding an individual's resistance? What still needs to happen in creating a context conducive to the change?

❒ Remember, change does not lead to an end point, but rather, it places us on a path toward new beginnings and greater possibilities. . . .

Important Disclaimer From Your Authors

To be certain that we do not leave the impression, by using the EveryWhere District setting, that the concepts and tools described and discussed in this volume are the only way/place to apply these ideas . . . the concepts and tools are equally applicable at the district and/or state level but also, importantly, in the grade-level team or secondary subject department settings and with single individuals. It is in these small environments that much learning for change and its implementation are considered and acted on.

Finale

The above graphic is a wordle that was created from all of the text in the finale.

Source: www.wordle.net/

If everything has gone as planned, most teachers and most schools will now be fully engaged with use of the new program. Most will be across the Implementation Bridge. An important number of research-based constructs, tools, and activities have been introduced and applied in the preceding parts of this book. For each construct a related professional development strategy (i.e., Learning Map) has been described. The foundation behind each strategy, or lesson, and its purpose is evidence based and built on years of research. In addition, all of the Learning Maps are grounded in real-world experiences with change. The suggested activities are based on ones that we have used. Our aim has been to offer action steps you can use to facilitate living change initiatives.

In this final part we offer some review, fill in some of the related conceptual and research background, and offer suggestions (and a few cautions) about what might happen next as you continue to facilitate various change initiatives.

IMPLEMENTING VS. SUSTAINING

The title for this section of our book, Finale, in many ways is misleading. Yes, it is the final section of this book. However, in terms of understanding and leading change initiatives, we have only walked and worked with you through the first 3 years of effort. The focus has been primarily on moving implementation well along. Achieving full implementation in schools typically takes 3 to 5 years. Transformational changes can take even longer. This time period is assuming that there has been continuing Concerns-Based change facilitation and that there has been leadership from principals, district administrators, and teachers. When there is no change leader or if there has not been continuing support, implementation can take more than 5 years or may never be accomplished.

We make a distinction between implementing and sustaining change. This book has been all about implementing a new practice, program, or process. All too often, once something is implemented, the leadership tends to lose interest in it. In too many organizations the top-level leaders have already shifted their attention to what comes next. If all of the implementation supports and attention on the emerging change move away, then it is likely that use of the new way will decline and perhaps disappear. Sustaining use of the new way requires that there be ongoing built-in system and leader supports.

Sustaining change could be the subject of another book. Sustainability is the current center of attention for researchers and most certainly an emerging concern for policymakers. There are too many cases where the heavy investments made in developing programs followed by serious efforts at scaling up and supporting new implementations have resulted in no lasting change, and 5 years later very little of "The Program" can be found. We know a lot about how to implement new programs and processes. We know a lot less about how to sustain them.

So although the title of this last section of the book is fittingly Finale, in terms of understanding and leading implementation processes, we are really at more of an intermission. The rest of the play/concert (sustaining) is yet to be performed.

GOING BACK TO THE BRIDGE

In the foreword of this book, the metaphor of the Implementation Bridge (Hall, 1999) was introduced. This metaphor was applied across the text to illustrate how the six implementation strategies and the three diagnostic dimensions of the Concerns-Based Adoption Model—Stages of Concern (SoC), Levels of Use (LoU), and Innovation Configurations (IC)—could be applied. The strategies and each of the three diagnostic dimensions represent distinct ways to describe the movement or journey of change and to measure how far across the bridge a change initiative has progressed.

Using the Implementation Bridge as metaphor for what is entailed in accomplishing major change probably makes more sense now. Each of the words and icons in the Implementation Bridge figure has special meaning. In addition, there are decades of research behind each. (I expect you are relieved that we did not regale you with reports of all the wonderful research and experiences with change efforts that are behind each of these ideas.) As a way of summarizing, the following are some of the important implications and reminders for you to keep in mind.

The Implementation Bridge

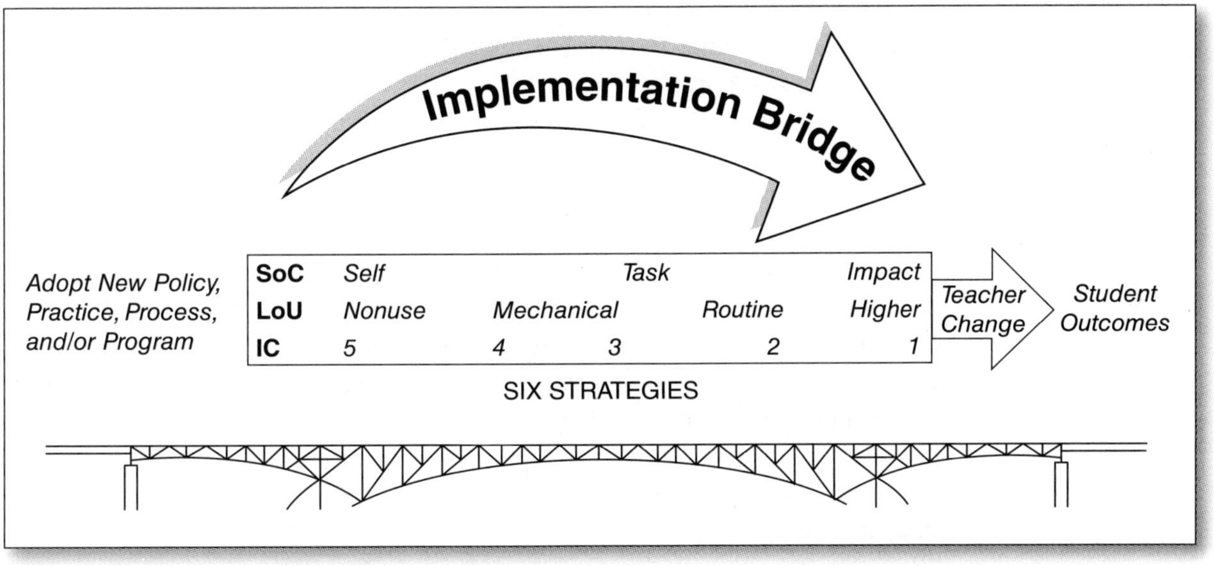

Source: Adapted by James Roussin from Hall, G. E. (1999). Using constructs and techniques from research to facilitate and assess implementation of an innovative mathematics curriculum. *Journal of Classroom Interaction, 34*(1), 1–8.

STAGES OF CONCERN

Many leaders appear reluctant to consider the *feelings and perceptions* that people have as they approach the Bridge and travel across it. However, failure to understand and appreciate the emotional part of change frequently leads to implementation failure. Resistance grows, and when leaders apply more pressure, some will back off the Bridge and a few will jump off the side.

A reality of change is that there is a personal side. As was described in Chapter 4, change facilitators need to design different supports in order to address different Stages of Concern. For example, when Stage 1: Informational concerns are intense, more information is needed (but not all in one big dump). When Stage 2: Personal concerns are intense, there needs to be a show of empathy and support (and not ignoring or failing to accept these concerns). In those special times when Impact concerns are intense, there should be celebrations and offering of extra supports. Sadly, often teachers (and administrators) with Impact concerns receive little or no encouragement. We get so focused on attending to those with Self and Task concerns that we forget about those who are already doing well.

LEVELS OF USE

In most research and program evaluation studies, use is considered to be dichotomous. Teachers use either the new program or the traditional program. The implicit assumption about change is that it is binary: present and absent, users and non-users. The field experiences behind development of the Concerns-Based Adoption Model led to understanding that there were a number of different behavior profiles of people in relation to what they were *doing*, or not doing, in regard to an innovation. Instead of use being dichotomous, we talk about *using*. We have identified a set of behavioral profiles to describe the different ways implementers engage with the change. Three non-use and five use levels have been operationally defined.

For those engaged in facilitating change processes, understanding each Level of Use is very important. Your change facilitating interventions need to be different for each LoU. For example, people at LoU 0: Non-use are taking no action to learn about the innovation. Interventions targeted toward them need to lead to their engaging with the idea of the change. Providing a little information at different points and in different ways over time to people at LoU I: Orientation contributes to their considering it. LoU II: Preparation people can be supported by assisting with organizing for first use. These are three different behavioral profiles, each of which is a non-user.

Levels of Use has important implications for program evaluation studies too. One of the most obvious is having implementers who are at LoU III: Mechanical identified as users.

By definition, LoU III people are novices. Subject samples for summative evaluation and treatment/control group studies should contain only implementers at LoU IVA: Routine, and possibly higher Levels of Use.

INNOVATION CONFIGURATIONS

It makes sense that just because teachers have been through introductory workshops and have the materials doesn't mean that all will fully implement the new way. Further, those that do engage with the innovation are likely to make it operational in different ways. Some will try to implement all the pieces and attend to every detail. Others will pull out the "good" ideas, and some will continue with what they had been doing. This is where the idea of different configurations came from. Although people will say they are "using" the new way, what they actually are doing can be very different. The only way to tell what is being done for sure is to make direct observations in each classroom. In some situations it may also be useful to do some related careful interviewing.

Developing an Innovation Configuration Map begins with identifying the critical components of a program or process. Then the possible variations of each component are described in ways that make each easily imagined and observable. Any combination of the component variations represents a distinct configuration.

As the IC variations evolve from *1* to *5*, there is a clear evolution away from the ideal to description of approaches that represent very little use of the new way, or none at all. You may have wondered about why the ideal Variation 1 is at the left on an IC map. One reason for this is so that the best practice will be the first to be read. As Stephen Covey would say, "Begin with the end in mind."

Having information about the current status of each IC map component is very important in planning what to do next. If many teachers are at Variation 4 of a certain component, then it might be useful to design a whole-staff meeting or a half-day professional development session targeted at that component. On the other hand, if it is only one or two classrooms operating with Variation 4, then the assistance should be individually targeted.

Another point—in recent years there has been increasing interest in *fidelity of implementation*. Curiously, in the past suggesting that implementers should strive to use an innovation the way the developers had envisioned was frowned upon. There was an expectation that each adopter would need to make adaptations in order to match the innovation to the local context. Now, with the heavy focus on evidence-based school improvement work, there are explicit expectations for implementers to operationalize the innovation with fidelity.

Documenting fidelity is another use of an IC map. The 1 variations should be descriptions of the developer's ideal. When an IC map is shared, implementers, leaders, and evaluators will have a common expectation for what implementation with fidelity looks like.

SoC, LoU, and IC are the three CBAM diagnostic dimensions. Each can be used by itself. In addition, any combination of the three can be used depending upon interest and need.

THE IMPLEMENTATION BRIDGE AND USING CBAM TO UNDERSTAND SUSTAINING CHANGE

One of the reasons for starting this Finale with the review and discussion of SoC, LoU, and IC was so that we can now use the constructs and metaphors to think about sustaining change. We have structured the discussion of this topic by asking several questions:

1. When will you decide that an implementation effort is a success?

What are the indicators you will look for? Is it when nearly all of the implementers are across the Implementation Bridge? What will SoC, LoU, and IC look like? What does it mean to be all the way across the Implementation Bridge? What are the minimums where sustaining use can be expected?

We use the three CBAM dimensions to define implementation success as follows:

- Implementation success in terms of SoC is when Self and Task concerns are low in intensity. This does not mean that Impact concerns are increased.
- Implementation success in terms of LoU is being beyond Level III: Mechanical. Implementers have to be at least at Level IVA: Routine.
- Implementation success in terms of IC is when minimally acceptable configurations are operational. With most IC maps this probably means Variation 3 is predominant, with few Variations 4, or 5, and there may be scatterings of Variations 1 and 2.

2. What does sustaining use entail?

Once implementers are across the bridge, they may continue using the innovation in the same way. However, it is more likely that there will be slippage. Some of the components that are more complicated and difficult to use start to be applied with less fidelity. As new teachers arrive, there is no training in how to use this innovation that has become well implemented and is now an "old" innovation. And when leaders change, there will often be changes in priorities and frequently the introduction of new programs and processes. Sustaining high-fidelity use of complex programs is not easy, or likely.

Sustainability begins with leadership maintaining attention on the new practice once there is implementation success. District Improvement Plans, as well as school improvement plans, need to continue to contain action steps related to the program. Sustaining also includes changes in some policies and building into budgets the necessary annual costs for resources, professional development, and coaching.

Another important part of sustaining can be the addition of related programs and processes. Once Self and Task concerns are reduced, it is possible to have more arousal of Impact concerns. Some teachers will now be concerned about finding and applying other ideas and activities that can further increase outcomes. This is the time to work on enhancements and perhaps add another innovation that compliments the first one.

3. Do SoC, LoU, and IC apply to groups and whole organizations, or just to individuals?

As you may have noticed, throughout this book we have not always drawn clear distinctions between the individual teacher's IC, LoU, or SoC and a whole staff's IC, LoU, or SoC. That is because each dimension can apply to individuals, groups, entire school staffs, and districts or statewide entities.

The measurements of SoC, LoU, and IC have been developed with sufficient research rigor so that they can be used confidently with individuals. The information about individuals can be aggregated to describe the current status of various groups such as the English Department or all first-grade teachers. The meaning of the constructs and interpretations will be the same. The individual assessments become the building blocks for looking at larger units.

IN CONCLUSION

The writing of this book has been a special opportunity for us to put into print some of what we have learned over the last four decades. During that time we have seen many very good programs come and go. Most were deserving of having a long life. In most cases their limited longevity was due to failures on the part of change leaders. Now our wisdom is such that we can usually predict the way a new change initiative will unfold. Our purpose in writing this book is to offer constructs, tools, and strategies based in our experiences that can be used to increase the likelihood of reaching implementation success.

In this book, through the story of EveryWhere School District and its change efforts, the authors have introduced the concept of an expanded, democratic model of users of the three CBAM Diagnostic Dimensions. They have promoted the *collaborating leader* and the *collegial facilitator* as a means to engage a wider base of participation in applying the power of these constructs to enhance and increase the possibilities for implementation success. They suggest broad sharing of CBAM knowledge and skills so that throughout the educational organization—district, school, grade-level team or subject matter department—Concerns-Based facilitation, support, and assistance can be learned, shared, and provided to all implementers. These facilitators will use the concepts, ideas, and theories of the CBAM Diagnostic Dimensions to stimulate conversation about change, as their learning communities gather to discuss, plan, and take action to support change projects.

Such dialogue and consideration of facilitation ideas and actions gives voice to the participants and invites them to grow in professional stature to become true professionals. This platform for the nurturing and development of professionalism is an important one where continuous adult learning, required for quality implementation, drives the conversation and the learning for all—where, ultimately, students are better served in their learning and successful achievement.

As the name suggests, the EveryWhere School District has many of the characteristics that we all know. The role of Bob Yurself is more difficult to assign. In some districts Bob, or Barbara, will be an assistant superintendent or director of professional development. In other settings he or she is a curriculum coordinator. Often, change efforts are not initiated or operated at the district level, but at the school level. In these cases the primary change facilitator will be a principal. In one district we know, very effective implementation leadership came from three master teachers who were assigned full time as districtwide staff developers and on-site coaches.

We hope that the ideas and suggestions presented in this book are informative and, most important, useful. Leading change and implementation efforts is not easy. It takes time and patience, but also persistence. Hearing the same Task concern question such as "How do I find time to make this activity work?" stated by four different teachers within 2 days can be frustrating or an irritation. Or it can be seen as a perfect example of why each person has to experience the Stages of Concern in his or her own way. With patience and the knowledge, skills, and perspectives from understanding change processes, it also can be seen with a sense of humor.

We wish you the best of success with your efforts to facilitate implementation processes. It is very important work. If the innovation is a good one, then making the time and taking the effort to facilitate implementers getting across the bridge is well worth it.

Also, be sure to have some fun along the way.

Gene E. Hall

References

Bandura, A. (1997). *Self-efficacy: The exercise of control.* New York, NY: Freeman.

Bransford, J. D., Barron, B., Pea, R., Meltzoff, A., Kuhl, P., Bell, P., . . . Sabelli, N. (2006). Foundations and opportunities for an interdisciplinary science of learning. In K. Sawyer (Ed.), *Cambridge handbook of the learning sciences* (pp. 19–34). New York, NY: Cambridge University Press.

Bridges, W. (2009). *Managing transitions: Making the most of change.* Reading, MA: Addison-Wesley.

Fixsen, D. L., Blase, K. A., Horner, R., & Sugai, G. (2009). Readiness for Change. *Scaling-Up Brief, 3.* Retrieved from http://sisep.fpg.unc.edu/resources/briefs/brief3readiness-for-change

Fullan, M. G. (2001). *The new meaning of educational change* (3rd ed.). New York, NY: Teachers College Press.

Fuller, F. F. (1969). Concerns of teachers: A developmental conceptualization. *American Educational Research Journal, 6,* 207–226.

George, A. A., Hall, G. E. & Stiegelbauer, S. M. (2006). *Measuring implementation in schools: The Stages of Concern Questionnaire.* Austin, TX: Southwest Educational Development Laboratory.

Hall, G. E. (1999). Using constructs and techniques from research to facilitate and assess implementation of an innovative mathematics curriculum. *Journal of Classroom Interaction, 34*(1), 1–8.

Hall, G. E., Dirksen, D. J., & George, A. A. (2006). *Measuring implementation in schools: Levels of use.* Austin, TX: Southwest Educational Development Laboratory.

Hall, G. E., & Hord, S. M. (2011). *Implementing change: Patterns, principles, and potholes* (3rd ed.). Upper Saddle River, NJ: Pearson.

Hall, G. E., Wallace, R. C., & Dossett, W. (1973). *A developmental conceptualization of the adoption process within educational institutions* (Report No. 3006). Austin: University of Texas at Austin, Research and Development Center for Teacher Education. (ERIC Document Reproduction Service No. ED095126)

Hord, S. M. (1992). *Facilitative leadership: The imperative for change.* Austin, TX: Southwest Educational Development Laboratory.

Hord, S. M. (1997). Professional learning communities: What are they and why are they important? *Issues . . . About Change, 6*(1).

Hord, S. M., Stiegelbauer, S. M., Hall, G. E., & George, A. A. (2006). *Measuring implementation in schools: Innovation configurations.* Austin, TX: Southwest Educational Development Laboratory.

Hord, S. M., & Tobia, E. F. (2012). *Reclaiming our teaching profession: The power of educators learning in community.* New York, NY: Teachers College Press.

Hoy, W. K., Tarter, C. J., & Woolfolk Hoy, A. (2006). Academic optimism of schools: A force for student achievement. *American Educational Research Journal, 43,* 425–446.

Joyce, B., & Showers, B. (2002). *Student achievement through staff development* (3rd ed.). Alexandria, VA: Association for Supervision and Curriculum Development.

Kennedy, J. (2012). Design learning that drives satisfaction. *The Leading Teacher, 7*(6) 2.

Kouzes, J., & Posner, B. (1995). *The leadership challenge: How to keep getting extraordinary things done in organizations.* San Francisco, CA: Jossey-Bass.

MetLife. (2011). *The MetLife survey of the American teacher: Teachers, parents and the economy.* New York, NY: Author.

Pink, D. (2010). *Drive: The surprising truth about what motivates us.* New York, NY: Penguin.

Senge, P. (1990). *The fifth discipline.* New York, NY: Doubleday-Dell.

Tobia, E. F., & Hord, S. M. (2002, March). Making the leap: Leadership, learning, and successful program implementation. *Instructional Leader.* Austin: Texas Elementary Principals and Supervisors Association.

U.S. Department of Education. (2010). *Learning: A model for the 21st century.* Retrieved from http://www.ed.gov/technology/draft-netp-2010/learning-model-21stcentury

Additional Resources for the Concerns-Based Adoption Model

What Is CBAM?

The Concerns-Based Adoption Model (CBAM) is a conceptual framework that describes, explains, and predicts probable teacher concerns and behaviors throughout the school change process. Introduction to the CBAM (video): http://www.sedl.org/cbam/videos.cgi?movie=Intro

The three principal diagnostic dimensions of CBAM are as follows:

Stages of Concern—Seven different stages of feelings and perceptions that educators experience when they are implementing a new program or practice Stages of Concern (video): http://www.sedl.org/cbam/videos.cgi?movie=SoC

Levels of Use—Eight behavioral profiles that describe a different set of actions and behaviors that educators engage in as they become more familiar with and more skilled in using an innovation or adopting a change Levels of Use (video): http://www.sedl.org/cbam/videos.cgi?movie=LoU

Innovation Configurations—Different ways an innovation may be implemented, shown along a continuum from ideal implementation or practice to least desirable practice Innovation Configurations (video): http://www.sedl.org/cbam/videos.cgi?movie=IC

CBAM TOOLS

The Southwest Educational Development Laboratory's (SEDL) CBAM tools and training will help evaluators, researchers, and administrators who are charged with measuring the implementation of a new practice in school settings.

Measuring Implementation in Schools: The Stages of Concern Questionnaire

George, A. A., Hall, G. E., & Stiegelbauer, S. M. (2006)

New to this edition is a set of files that includes a Microsoft Word version of the questionnaire and scoring sheets, as well as a scoring program in Excel and SAS formats.

View the SEDL Store listing for this product (CBAM-17):

http://www.sedl.org/pubs/catalog/items/cbam17.html

Measuring Implementation in Schools: Levels of Use

Hall, G. E., Dirksen, D. J., & George, A. A. (2006)

The manual discusses using a focused interview and the LoU chart to assess use.

View the SEDL Store listing for this product (CBAM-18):

http://www.sedl.org/pubs/catalog/items/cbam18.html

Measuring Implementation in Schools: Innovation Configurations

Hord, S. M., Stiegelbauer , S. M., Hall, G. E., & George, A. A. (2006)

This updated manual provides step-by-step instructions for developing a map, numerous examples, and insight into data collection and interpretation.

View the SEDL Store listing for this product (CBAM-19):

http://www.sedl.org/pubs/catalog/items/cbam19.html

To facilitate data collection and analysis, SEDL now offers an online version of the Stages of Concern Questionnaire: http://www.sedl.org/pubs/catalog/items/cbam21.html

Index